Canada's Forests at a Crossroads: An Assessment in the Year 2000

GLOBAL FOREST WATCH

Linking forests & people

www.globalforestwatch.org

A Global Forest Watch Canada Report prepared by:
Wynet Smith, Bryan Evans, Tim Wilson, Andrew Nikiforuk, Karen Baltgailis, Allyson Brady, Laurel Brewster, Tim Gray, Will Horter, Peter Lee, Robert Livernash, Claire McGlynn, Susan Minnemeyer, Aran O'Carroll, Alan Penn, Geoff Quaile, Mike Sawyer, Elizabeth Selig, and Dirk Bryant

Partners
Alan Appleby, Allyson Brady, Tim Gray, Sam Gunsch, Julian Holenstein, Will Horter, Henri Jacob, Peter Lee, Garry McLean, Aran O'Carroll, Geoff Quaile, Don Sullivan, and Ken Young

Global Forest Watch is an Initiative of the World Resources Institute

World Resources Institute
Washington, DC
www.wri.org

Managing Editors:
Wynet Smith (World Resources Institute)
Peter Lee (Global Forest Watch Canada)

Designed by:
Papyrus Design & Marketing, Washington, DC

Production by:
Beacon Hill Communications Group Inc., Victoria, BC

Cover photographs (left to right):
© Garth Lenz; Haida Brave log barge, Strait of Juan de Fuca, BC, Canada; 1992.
©Wynet Smith; Fireweed, Yukon Territory, Canada; 1997.
©Greenpeace/Ivan Hunter; Coastline, South of Klemtu, Great Bear Rainforest, BC, Canada; 1998.
Background: ©Parks Canada; Maple leaves, La Mauricie National Park, Québec

Each World Resources Institute report
represents a timely, scholarly treatment of a subject of public concern.
WRI takes responsibility for choosing the study topics
and guaranteeing its authors and researchers freedom of inquiry.
It also solicits and responds to guidance of advisory panels and expert reviewers.
Unless otherwise stated, however,
all the interpretations and findings set forth in WRI publications
are those of the authors.

ISBN 1-56973-440-2
Library of Congress Catalog Card No. 00-101076
Printed in Canada on Recycled Paper

CONTENTS

FIGURES

TABLES

BOXES

TECHNICAL NOTES

MAPS

ACKNOWLEDGMENTS

Global Forest Watch Canada wishes to thank the following individuals for their support, contributions, and comments on the report:

Frank Ahern (Canadian Centre for Remote Sensing); Jim Ball; David Boyd; Laurel Brewster (BC Forest Watch); Linda Coady and Jean-Pierre Martel (Weyerhauser); Peter Duinker (Dalhousie University); Arlin Hackman (World Wildlife Fund Canada); Peter Hall (Natural Resources Canada); Mark Harmon (University of Oregon); Bill Hunter (Alberta Pacific Forest Industries, Inc.); Tony Iacobelli (World Wildlife Fund Canada); Trevor Jones, Eric Kasischke (Geography Department, University of Maryland), Ajit Krishnaswamy (International Institute for Sustainable Development), Claude Leger (Canadian Forest Service); Stuart Loomis; Patricia Marchak (University of Victoria); Giselle Marcotte (Saskatchewan Intergovernmental and Aboriginal Affairs); David Morel and Larry Pederson (British Columbia Ministry of Forests); Linda Nowlan (West Coast Environmental Law Centre); Dr. William Pruit (University of Manitoba); Dr. B. Purdy; Dr. Stan Rowe; Monique Ross (Canadian Institute for Resources Law, University of Calgary); Glen Semenchuk (Federation of Alberta Naturalists); Peggy Smith (National Aboriginal Forestry Association); Brad Stelfox (Forum Consulting Ltd.); Martin Von Mirbach (Centre for Forest and Environmental Studies); and Alan Young (Environmental Mining Council of British Columbia).

Many World Resources Institute staff also reviewed, provided input, or helped prepare the report: Matt Arnold, Hyacinth Billings, Dirk Bryant, Susan Buzby, Tony Janetos, Nels Johnson, Jonathan Lash, Lars Laestadius, Peter Leimgruber, Robert Livernash, Andrew Malk, Susan Minnemeyer, Marta Miranda, Cathy Plume, Mark Rowheder, Elizabeth Selig, Nigel Sizer, and Wynet Smith.

FOREWORD

This report—*Canada's Forests at a Crossroads: An Assessment in the Year 2000*—is one of the first products of Global Forest Watch (GFW), a remarkable new alliance that was launched by the World Resources Institute in 1997 with the help of nongovernmental organizations (NGOs) and local leaders from forested countries around the world. GFW links satellite imagery with on-the-ground investigation by local groups to assemble powerful information about the risks to the world's great forests. It then uses the Internet to make the information widely available.

Technological innovation can change the way we manage and protect our forests and environment. First, technology provides us with the tools we need to get accurate and up-to-date information about forests and other ecosystems, a prerequisite for informed decisionmaking. Second, technology provides the means to make this information available to all those with a stake in natural resources. Providing this information will help ensure that resources are managed for the common good.

Until recently, there was little systematic knowledge about the condition of the world's forests. It was impossible to say how much forest had been lost and how much remained as frontier forest–large, intact, and fully functioning natural ecosystems. Frontier forests are important to human well being. Forests help to slow global warming because they store vast quantities of carbon. They control flooding, purify water, and cycle nutrients and soil, ultimately influencing food production for billions of people. And they house an incredible array of living organisms, which provide a foundation for the resilience of natural systems and the genetic material for valuable new products.

In 1997, WRI and our partners collaborated with scientists and local experts around the world to map out remaining frontier forests and areas that had been cleared in past generations. This work could not have happened without new information tools at our disposal: geographic information systems to store and analyze data; access to maps derived from satellite images; and the Internet to share drafts and exchange results with our collaborators. Our report, *The Last Frontier Forests*, established that just 20 percent of the frontier forests that once blanketed the earth remain today. Much of what is left is under intense development pressure, primarily from logging and other extractive use.

Existing forest monitoring efforts have primarily been confined to tracking deforestation and forest degradation after it has happened. This work has limited use for management decisions, because once an area has been cleared or degraded, it is frequently too late to do anything about it. To fill this information gap, GFW seeks to provide early warning data on forest development and on the environmental and economic tradeoffs that development entails. GFW empowers local organizations to monitor and report on their forests, assisting growing civil society institutions to gain access to remote sensing technology and the power of the Internet. These organizations are connected to a worldwide network of partners bound together by a commitment to accurate information and open dialogue about forest management. Grounded in the idea that more public information helps create better outcomes, GFW aims to become an independent source of timely and practical information about where forests are being developed, by whom, and how.

Through this report, our Canadian partners have documented that logging, mining, and other development now occurs throughout much of Canada's forests. The most diverse and productive forest ecosystems—including temperate coastal rainforests in British Columbia—have undergone widespread fragmentation by roads and other access routes and have the bulk of their area under logging tenures.

Canada is at a crossroads. Public sentiment and new government policies indicate an increasing commitment to managing forests not just for timber, but also for wildlife, recreational uses, cultural values, and other ecosystem services. However, changes on the ground have been slow. The interests of First Nations and Métis, who hold long-standing claims to large tracts of forest, are still underrepresented in forest management decisionmaking. Most logging still occurs within primary and old-growth forests, while tenures increasingly extend into far-northern, ecologically sensitive regions. Deep budget cuts have hampered government capacity to implement and enforce new management regulations. Many of these responsibilities are being shifted to the private sector. Due to a wave of consolidations, vast areas of Canadian forest are managed by a handful of timber companies. These corporations can play an influential role in promoting forest management policies that factor in social and environmental values associated with forests.

Our most surprising finding is that much of the data needed to help make these management transitions is not available, publicly accessible, or affordable for many groups. We believe that additional public information will promote accountability and informed dialogue. Information creates understanding, which can be the basis for trust. Transparency builds incentives for the implementation of commitments made to manage and protect the world's forests, which would help slow forest degradation around the world.

GFW seeks to make information available rapidly to an ever wider audience by providing forest information and maps on line and developing a state-of-the-art Website (www.globalforestwatch.org) to post results from its multiple field activities in Cameroon, Canada, Chile, Gabon, Indonesia, Russia, and Venezuela. Reports, maps, and other transparent information will be available for download. Anyone with access to the Internet can consult GFW data. Furthermore, they can contribute by providing information or views directly on-line. We hope that the array of products and activities will contribute to a more constructive dialogue between forest managers and users at the local, national, and international level.

We would like to thank the following donors for overall support of Global Forest Watch activities: AVINA, the Department of International Development (DFID), UK, IKEA, the Netherlands Ministry of Foreign Affairs, and the Turner Foundation.

In addition, we would also like to thank the Richard and Rhoda Goldman Fund and the Bullitt Foundation for their support of Global Forest Watch Canada activities.

Jonathan Lash
President
World Resources Institute

KEY FINDINGS

In recent years, international attention has increasingly focused on the rapid conversion and degradation of the world's tropical forests. Yet half of the remaining large tracts of natural forest are found in northern (or boreal) regions.

This report provides a first look at the scale and magnitude of development within Canada, one of the world's major repositories of northern forests. Canada is home to over a third of the world's boreal forest and a tenth of total global forest cover. Largely unsuited to agriculture, these forests have escaped widespread conversion to farmland and ranches—key threats in tropical regions. This northern frontier is rapidly being opened up for its timber, energy, and mineral resources. Logging is the dominant activity and a key sector in Canada's economy; the forest industry generated over $68 billion in sales and $11 billion in wages in 1996.

Global Forest Watch (GFW) Canada—a network of regional and local environmental groups and First Nations—has set out to answer the four basic questions GFW addresses worldwide. What large-scale development is occurring in forests, and where? What environmental impacts and economic benefits does this development entail? Who are the key actors engaged in these activities? Are these activities compatible with legislation set out to promote forest stewardship? In this report, we present preliminary results, drawing largely on available data, but including new analysis derived by combining spatial (mapped) datasets through the use of Geographic Information Systems (GIS).

Canada's forests are managed predominantly for timber. However, the Canadian public values forests primarily for nontimber uses.

Some 94 percent of Canada's forests are held in the public trust by federal and provincial governments. Polling data indicate Canadians most value forests for nontimber uses: for species habitat; for ecosystem services such as watershed protection and carbon storage; and for intrinsic wilderness value. However:

- 52 percent of forests are managed as logging tenures. In contrast, less than 8 percent of Canada's forests are fully protected, although many new parks and reserves have been established in recent years.
- Of 10 major forest types, 6 have at least two thirds of their area allocated as logging tenure.
- Canada maintains its lead as the world's biggest timber exporter through logging of old-growth and primary forests, which account for 90 percent of the harvest.
- Clearcuts make up over 80 percent of annual harvested area. Although economically efficient, clearcutting results in quite different disturbance patterns than fires and other natural processes. The ratio of clearcut area to the area using partial harvest systems has remained unchanged over the last two decades.
- 95 percent of all major forested watersheds include roads, mines, settlements, and other developments. These pose unquantified threats to watershed protection functions, carbon storage, and other ecosystem services provided by forests.

Canada's most species-rich and productive forests have been extensively modified by development activities.

- Over half of the forests in 7 of Canada's 10 major forest regions have been fragmented by roads and other access routes.
- About three-fifths of the eastern Carolinian forests and the Aspen forests bordering the prairies have been converted to agricultural and residential land.
- Coastal forests of British Colombia—home to one fifth of the world's remaining temperate rainforest and noted for exceptional biodiversity—are under widespread development pressure. Over 80 percent of this forest has been allocated to logging companies (through tenure areas managed for timber harvest, which include extensive tracts of forest not destined for cutting). Nearly half the forest is fragmented by roads and access routes in blocks less than 200 km^2 in size.

Under current management practices, harvesting rates appear unsustainable over the long term.

Only 1 million of Canada's 235 million hectares of commercial forest land are cut annually. However, this figure—because it factors in marginally productive lands and does not account for extensive areas affected by fires and other natural disturbance—understates the implications of current harvest rates.

- 50 percent of tenured areas face productivity limitations due to climate, topography, and other factors.

- Harvest quotas are often set above long-term sustainable yields. For example, in British Colombia, the leading provincial producer of timber, 90 percent of lands managed for harvest (timber supply areas) are logged above long-term sustainable levels set by the government.

A handful of companies now manage much of Canada's forest.

Industry consolidation has resulted in the concentration of vast areas of forest in the hands of a few companies. These corporations—because of the revenues and jobs they control—are in a position to significantly influence provincial forest policies.

- 13 companies have holdings at least the size of Switzerland, accounting for over 48 percent of Canada's forest tenure areas.

- About 80 percent of Canada's First Nations and Métis live on reserves and communities in boreal or temperate forests. Although aboriginals hold extensive and longstanding claims to Canada's forests (many unresolved), these areas are largely allocated to and managed by the private sector. Management for timber production often conflicts with First Nations' rights and traditional holistic values toward forests.

Development increasingly extends into Canada's northernmost forests.

Popularly viewed as an endless expanse of wilderness, the Boreal and Taiga (transition forests at the edge of the tundra) Forest Regions encompass almost 1.9 million km^2 in unfragmented blocks at least 10,000 km^2 in size. However, these forests are being opened up, primarily for energy and mineral resources, but also for timber. The potential impacts of these activities are unknown. Canada's northernmost forests are particularly sensitive to development, in part because short growing seasons and fragile soils limit vegetation regrowth.

- At least 300 hydro dams, 80 active mines, and over 1,400 settlements are found in the Boreal and Taiga Forest Regions.

- 30 percent of the Boreal Forest Region is within a kilometer of a road or access route.

- Logging tenures now extend into Canada's northernmost and most ecologically sensitive forests. Almost 50 percent of the Boreal Forest Region is under tenure.

Increasingly, Canada is promoting sustainable forest management policies. However, implementation remains a problem.

This report includes an impressive list of new policies and initiatives established by Canadian governments to promote forest stewardship. It provides incomplete information—derived largely from independent review panels—on progress made in implementing these policies.

Information and data collected by Global Forest Watch Canada partners indicate declining public oversight over forests. Widespread cuts in government budgets and staffing have resulted in forest planning, management, and enforcement responsibilities being shifted increasingly to industry.

Lack of publicly available forest information hinders accountability and informed decisionmaking.

- As a result of cost-recovery policies, government datasets are often prohibitively expensive to noncommercial users.

- National datasets on productivity limitations, land ownership, aboriginal forest use, threatened species distributions, and compliance with management laws are either outdated or not systematically collected.

- There is no systematic monitoring of changes in forest condition—for example, where primary forests are being converted to secondary growth, which is useful for gauging the environmental tradeoffs development entails.

Global Forest Watch Canada seeks to work with government, industry, and other groups to make such data widely available and to promote informed decisionmaking in favor of long-term planning and management driven by public interests.

SECTION 1. INTRODUCTION

Unlike most other developed nations, Canada's economy and national identity are closely tied to its vast forest resources. Canada's national territory includes about 10 percent of the world's forests, 35 percent of the world's boreal forests, and 20 percent of the world's temperate rainforest.[1] Canada contains about one fourth of the Earth's remaining frontier forest—the large, relatively undisturbed forest areas with sufficient area to maintain all of their native biodiversity.[2]

Canada is one of only eight countries with an opportunity to maintain most of its frontier forest intact.[3] Protecting these resources or ensuring that development proceeds on a sustainable basis should be a vital priority, both for Canada and for the rest of the world.

The reasons are well known. As evidence of climate change continues to mount, so too will the importance of Canada's forests as storehouses of carbon. North American forest ecosystems store a significant proportion of the global total of biotic carbon. At the same time, the impacts of climate change could pose an unprecedented threat to Canada's forest resources.

Canada's forests are also home to a remarkable diversity of plants, animals, and microorganisms. They are a refuge for woodland caribou, grizzly bear, grey wolf, and other large mammals that once ranged widely across the continent. There are an estimated 140,000 species in Canada, only half of which are classified. About two thirds of these species are found in forests or are dependent on forest habitat. New species continue to be discovered; scientists recently identified 60 new insect species in the canopies of old-growth forests in British Columbia.[4]

Canada's boreal forests play a particularly vital ecological role, both as storehouses of biodiversity and as a vast reservoir of freshwater and carbon. In recent decades, these forests have been the focus of major hydroelectric developments, oil and gas exploration, and more recently, logging. The boreal forests will face unprecedented stresses in the coming decades, both from direct development pressures and from the effects of global climate change.

Canada's forests also play an important role in the nation's economy and society. The forest industry generated over $68 billion (US $47 billion) in total sales in 1996. In addition, it directly employed over 350,000 Canadians in 1998. Canada continues to be the world's largest exporter of forest products. Canada also is one of the world's top mineral producers, with almost 300 metal, nonmetal, and coal mines. Oil and gas exploration and development is a major activity in western Canada. Other commercial activities include hunting, trapping, fishing, and tourism.

The benefits of these economic activities are important globally as well. Canadian forests provide wood products to many areas of the world, including the United States, Europe, and Japan. Houses in the U.S. are built from Canadian wood and newspapers are printed on paper from Canada. Oil and gas are exported. Minerals extracted from within Canadian forests provide essential raw materials and products to many countries.

Canada's forests are a rich cultural legacy and source of sustenance for First Nations and Métis. Almost 80 percent of Canada's 1 million aboriginal people live in communities throughout Canada's forest regions. Reconciling the interests of Canada's aboriginal peoples as forests are developed will be a key challenge for the nation in the coming decades.

It is estimated that about one fifth of Canada's remaining frontier forests are directly threatened by logging, mining, agricultural clearing, and other human activities. These activities will eventually degrade forest ecosystems and result in large-scale changes in the forests' age and structure.[5] This report provides more information on the details and nature of these activities. For example, roads, urban and rural expansion, and industrial activities have now fragmented most of Canada's southerly forests. Forests continue to be converted at a rate of about 55,000 to 80,500 hectares per year, and logging is occurring on about 1 million hectares per year. While over half of forests are under some form of tenure for wood production, analysis in this report reveals that extensive forest areas have ecological constraints to commercial forestry. (Tenures are license agreements between provinces and companies that grant the companies rights to cut timber but not rights to other resources. A more detailed explanation of tenures is provided in Section 3.)

Key stakeholders in the management of Canada's forests include the federal, provincial, and territorial governments; the private sector; First Nations and Métis; forest-dependent communities and workers; and private individuals. Under Canada's Constitution, provincial governments are primarily responsible for forest management and other resources. Each province has legislation, regulations, policies, and programs governing the allocation of forest use, harvesting rights, and management responsibilities. The federal government is primarily responsible for Canada's participation in numerous national and international forest agreements and in collecting information and reporting on forests on a national basis. Numerous changes to Canadian laws and policies in the last decade may signal a shift in forest management priorities.

In this report, Global Forest Watch Canada documents—through a series of maps and indicators—the extent of industrial activity in Canada's forests, along with key actors (governments, companies, and specific groups) engaged in this development. We provide a more in-depth look at the costs and benefits of the logging industry, which is the most widespread development activity occurring within the forests. We provide an overview of key policies, initiatives, and legislation that are designed to help promote forest stewardship and progress on implementing these initiatives.

We attempt to answer four key questions addressed by Global Forest Watch partners world-wide:

- What large-scale developments are occurring and where?
- What environmental trade-offs does this development entail?
- Who is engaged in these activities?
- Do these activities comply with existing international, national, and local laws, as well as standards and agreements?

The three main sections of the report are:

- Indicators of Forest Condition and Change, which presents data on large-scale development and potential environmental trade-offs;
- The Forest Industry, which focuses on the extent and location of logging, its beneficiaries, and its sustainability; and
- Commitments and Legislation, which describes legislation and initiatives and progress in their implemention.

Key findings are highlighted. Technical notes in the report provide more information on the data, methods, and caveats about the findings. Appendixes 1 and 2 provide further information on methods and datasets used to generate the results in the report and on the review process.

Data: Methods and Key Challenges

This report contains information from published sources that was compiled to present a national overview of key forest trends. Wherever possible, Canadian government data and sources were used as the basis for original mapping and analysis in this report. We also present new map-based analysis that was developed through the use of Geographic Information Systems (GIS). GIS combines map data to illuminate relationships—for example, the degree to which roads are fragmenting forests—not otherwise captured by maps with a single dataset.

Forest industry data are derived from a few sources. We have used the *National Forestry Database Program* (compiled every year by the Canadian Council of Forest Ministers), the *Canadian Forest Inventory* (compiled every 5 years by the Canadian Forest Service), and related publications for much of our forest information. *Statistics Canada* and the *National Forestry Database Program* are the source of much of the data on the economics and employment aspects of the forestry and other resource industries, although we have also used and referenced other sources. We have used the 1995 *Land Cover of Canada* map, which is available from the Canadian Centre for Remote Sensing of Natural Resources Canada. Other key data sources have come from Environment Canada and Agriculture Canada. GFW Canada also compiled a number of important provincial datasets for use in these analyses.

Some of the key challenges we faced in preparing this report relate to the state of datasets in Canada. Canada has a vast amount of digital data available. Interpreting the data is often difficult, however, in part because the datasets are compiled by a variety of jurisdictions, are available in a variety of formats, and are not consistent nationally. GFW Canada also was constrained by the prohibitive costs of purchasing and assembling these datasets.

In some cases, we have used national datasets that lose local information. While this provides challenges for the regional level of analysis, we did this in order to provide a national picture. As a next step, we intend to develop regional levels of analysis that will be based on more detailed, finer resolution data. This report includes some examples of the type of work that can be extended to other regions to build a more complete picture of forest trends.

A detailed description of source data and mapping methods is provided in Appendix 1. All sources are cited and a World Wide Web reference is provided wherever it is available.

Unless otherwise noted, all dollar amounts are Canadian dollars. At the time of publication, 1 Canadian dollar was equal to US$0.69 (US$1=Canadian $1.46).

More than 30 people reviewed the draft report, including representatives of federal and provincial governments, the private sector, academia, and nongovernmental environmental organizations. As noted in Appendix 2, this final report reflects changes made in light of that review process.

SECTION 2. INDICATORS OF FOREST CONDITION AND CHANGE

INTRODUCTION

In this section, we provide an overview of Canada's forest and the human activities—such as clearing for farming and settlements, logging, mining and other development—that influence forest extent (area) and condition. This responds to GFW's mandate to 1) track existing and planned development activities and 2) provide data on forest ecosystems to highlight the environmental and economic trade-offs that development options entail.

Forest condition, which is sometimes equated with forest health, is a relative term that depends on how people value forests. For example, intensively managed forest plantations emphasize timber production over nontimber values such as recreational opportunities and wildlife habitat. From the perspective of a conservation biologist, forest plantations are considered degraded forests. Old-growth forests, noted for standing dead trees and fallen logs, provide key habitat for certain species and store vast amounts of carbon, but to loggers they represent wasted timber resources.

On what basis should forest condition be assessed? Given that the vast majority of Canada's forests are publicly owned, public opinion polls can provide an indication of the public's priorities for Canada's forests. A 1997 survey commissioned by Natural Resources Canada reported that, while Canadians understood the importance of benefits from the forest industry, industrial use was not what they

Summary of Overall Forest Condition and Change Trends

Analysis of forest development trends in Canada indicates that relatively little forest has been cleared in total. Much of what remains outside of the northern Boreal and Taiga Forest Regions, however, is fragmented by roads and other access routes or is close to mines, settlements, and other development. Forests in the southern half of the country—those that are the most productive and species-rich—have been extensively modified (as measured by access and conversion). Of the 10 major forest types found within Canada, 2 have lost about 60 percent of their forest cover. Seven of these 10 have more than half their remaining forest area fragmented by access routes.

These trends are likely to correspond to a significant loss of wilderness values associated with forests and the loss of habitat for species, particularly those that are highly sensitive to human disturbance or that live in forests with most of Canada's biodiversity. This factor potentially impacts such ecosystem services as maintenance of water quality and provision of habitat for aquatic species. In the majority of Canada's forested watersheds, development is found within at least 25 percent of the watershed area.

Over half of Canada's forest area still remains in very large blocks of unfragmented (unaccessed) forest, almost entirely in the northern Boreal and Taiga Forest Regions. These areas present significant opportunities for maintaining biodiversity and globally important ecosystem services, such as carbon storage.

Development now extends into Canada's northernmost forests. This includes over 80 active mines, over 300 hydroelectric dams, and over 1,400 settlements in the Boreal and Taiga Forest Regions. Large areas of the Boreal Forest Region are also tenured for timber production. Our analysis identifies those areas most likely to be affected by development, but does not indicate the actual magnitude of impacts.

valued most about their forests. *(See Figure 1.)* Canadians placed the greatest importance on environmental and ecological benefits derived from forests, such as the role they play in protecting the nation's water, air, and soil. *(See Box 1 for a description of ecosystem services derived from forests.)* Next in order of preference were forest values pertaining to wildlife habitat, followed by wilderness preservation and economic benefits. Recreational values were rated last.[6]

We set out to answer, to the extent possible based on existing data, four specific questions:

1. How much forest does Canada have?
2. What types of forest are found in Canada? How are they unique?
3. What industrial activities are occurring within Canada's forests?
4. What potential trade-offs (impacts) do these activities have on biodiversity, wilderness values, and other ecosystem services derived from forests?

Questions 1 and 2 address forest cover trends, or the nature and extent of Canada's forest assets. Questions 3 and 4 focus on assessing forest condition, especially regarding the values important to Canadians.

Using a variety of existing datasets and maps, we have derived eight indicators, or measurable factors, which can be used to answer these questions and set the framework for long-term monitoring of Canada's forests. These are described below.

Figure 1. Importance Placed on Forest Values by Canadians, 1997.

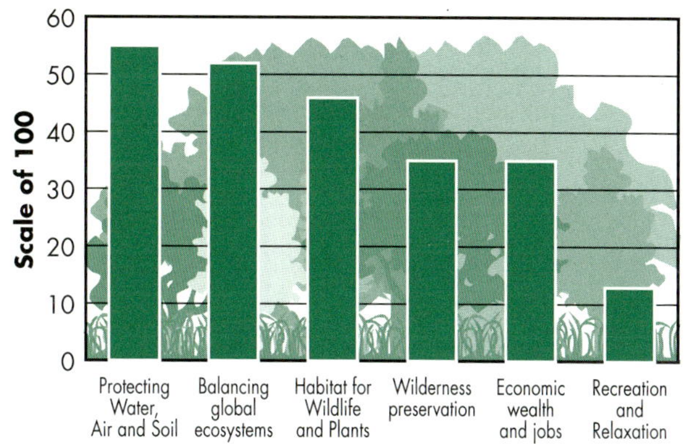

Source: Corporate Research Associates Inc., *Tracking Survey of Canadians Attitudes Toward National Resource Issues, 1997* (Ottawa: Natural Resources Canada, 1997), p. 36.

Forest Indicators

Forest Cover

Indicator 1 – Forest extent (amount): This shows remaining forest area, and provides a baseline (or starting point) for assessing changes to Canada's forests.

Indicator 2 – Forest regions (types): A national map showing the major types of forest found in Canada.

Forest Condition

Indicator 3 – Cumulative forest development: Statistics on the overall amount of development (mines, hydroelectric development, roads, settlements, and forest tenures) occurring in Canada's forests can be used to estimate the extent and location of specific types of development activities.

Indicator 4 – Watershed development: A national map showing the degree to which major forest watersheds have been developed. Watersheds are ranked according to the proportion of watershed area close to roads, mines, hydroelectric development, and settlements. This map and associated statistics are a measure of condition in terms of some potential impact on a specific set of environmental services—those associated with the role forests play in watershed protection, such as maintenance of water quality and of habitat for aquatic species.

Indicator 5 – Forest conversion (clearing): Farmland and settled areas within major forest regions provide an estimate of where forests have been cleared to date for these types of uses. This estimate is a measure of change in forest extent and potential impacts on broader ecosystem services such as carbon sequestration.

Indicator 6 – Accessed Forest: Roads and other access routes in current forest area show where forests have been fragmented to date. This indicator provides a measure of change in extent and condition and possible loss of biodiversity and ecosystem services such as wildlife habitat and carbon storage.

Indicator 7 – Unfragmented forest: Large areas of unfragmented forests (where no roads or other access routes exist) show the extent of forest remaining that offers prime wilderness and wildlife habitat (important conditions for biodiversity and wilderness values).

Indicator 8 – Forest-dwelling species at risk: Data on the number of forest-dwelling species at risk provide a direct measure of threats to Canada's biodiversity.

These measures are rough indicators of the possible loss of ecosystem services. They are not actual values forfeited as a result of development. Such values are impossible to quantify nationally with currently available data. What these indicators show is the total area and location of potentially impacted forests. These indicators are described further in short technical notes sprinkled through the text and in Appendix 1.

BOX 1 What Are Ecosystem Services?

Forest ecosystems provide humans with an array of beneficial "services." These services come from ecological functions such as nutrient and water cycling, carbon sequestration, and waste decomposition. A few of the more prominent forest ecosystem services include:

- Moderation of water flow extremes. Forest vegetation slows the flow of water through the landscape. Flooding is less severe in intact, forested watersheds during high rainfall or rapid snowmelt events. The cool, shaded forest floor and the slow movement of water through forest soils also extends the flow of water during severe droughts. Economic damages from flooding and drought are often much higher in watersheds with degraded forests than in watersheds with natural or sustainably managed forests.

- Water purification. Intact, forested watersheds protect water quality because vegetation prevents erosion and forest soils filter out impurities.

- Moderation of global and regional climate. At the global level, forests regulate climate by sequestering carbon dioxide from the atmosphere. At the regional level, forest vegetation recycles some rainfall back into the atmosphere through evapotranspiration, thus helping to refuel rainclouds and maintain regional precipitation levels. Deforestation, mostly in the tropics, now contributes 20 percent of greenhouse gas emissions.

- Maintenance of genetic diversity. Forest ecosystems are home to a majority of the world's terrestrial species. This diversity gives rise to a wide variety of useful products, including pharmaceutical compounds such as taxol from Pacific yew trees, which is effective against ovarian cancer. The loss of species and genetic diversity deprives the rapidly growing biotechnology industry of potentially useful new compounds that could help conquer disease.

- Provision of recreational and cultural opportunities. Forest ecosystems provide people with some of their favorite environments for recreation. In some forest areas, the economic value of hiking, camping, fishing, hunting, wildlife viewing, and other recreational activities can rival or exceed returns from timber or other consumptive uses. Forest areas often represent important cultural values to people, especially indigenous peoples.

FOREST COVER

Indicator 1: Forest Extent

After Russia and Brazil, Canada has more forest than any other country

Of Canada's 922 million hectares of land area, 417.6 million hectares are forested.[7] This total includes areas clearcut and left to regrow, along with forests not considered commercially productive.[8] Forests currently cover 45 percent of Canada's landmass.[9] One third of Canada—over 340 million hectares—encompasses large intact blocks of natural "frontier" forest and forest-tundra.[10]

Over half—244.6 million hectares—of Canada's forest area is considered commercial or "timber productive" forest, which is suitable for timber harvest. Of this amount, 235 million hectares are actually available for commercial use. Nine million hectares are not available for harvesting.[11]

Indicator 2: Forest Types

Canada has 10 different major forest types/ecosystems

Canada is home to several globally important forest types. Map 1 shows the 10 major forest types (Forest Regions) found in the country. *(See Technical Note 1.)* The forest types each have characteristic tree species. Their extent (total area and percent of land area) varies widely. *(See Table 1.)*

Over one third of the world's boreal forest is found in Canada

Boreal forests predominate in Canada and actually consist of three different areas: Boreal, Taiga, and Aspen Parkland Forest Regions. The last two are a mix of trees and open land—transition ecosystems between boreal and tundra in the north, and boreal and grassland in the south. Rather than grouping them together, all three categories are treated separately in this report. We use the term boreal forest to refer to all three categories and Boreal Forest Region to identify the more restricted area.

Canada's boreal forests are of global importance. They encompass much of the world's remaining large tracts of intact forest; provide feeding and breeding grounds for key migratory species; store vast quantities of carbon; and provide a host of other ecosystem services. Global warming and its impacts on Canada's forests—especially the boreal—is the subject of considerable discussion in Canada. *(See Box 2.)*

One fifth of the world's temperate rainforest is found in British Columbia

The Coast Forest Region, found in British Columbia, is one of the largest remaining areas of coastal temperate rainforest in the world.[12] Originally covering 30 to 40 million hectares, researchers estimate that 17.3 million hectares (or 56 percent) have been logged or converted to nonforest uses worldwide.[13] No intact, unlogged watersheds of any size remain in the continental United States; only British Columbia and Alaska have some large, undeveloped tracts remaining.[14] Coastal temperate rainforests contain high levels of biodiversity, much of which has yet to be described and named. They are among the most complex and dynamic systems on earth and are considered rare on a global scale.[15]

A 1990 assessment of the status of British Columbia's coastal temperate rainforests reported that two thirds of the 354 primary coastal temperate watersheds of British Columbia were developed and that less than one fifth remained undisturbed.[16] A majority of the undeveloped watersheds are located on the mid-coast and north coast of British Columbia, and include the Kitlope in Northern British Columbia. In 1994, the British Columbia government (in cooperation with the Haisla First Nation and West Fraser Mills Ltd., which surrendered its tenure without compensation) protected the Kitlope River, which encompasses the world's largest intact tract of coastal temperate rainforest.[17]

British Columbia's mainland coast is home to one of the largest remaining tracts of frontier temperate rainforest in the world—an area referred to as the Great Bear Rainforest by some groups.[18] Here, significant opportunities remain to protect viable, representative examples of temperate rainforests and fully intact watersheds on the central coast, along with healthy populations of grizzly bear, salmon, wolf, wolverine, and a rare white-furred variation of the black bear called the Spirit Bear. A government-led planning process is under way to develop a land use plan for this area. The effort includes the major forest licensees (International Forest Products, West Fraser, and Western Forest Products), First Nations, conservationists, and other stakeholders.[19]

Table 1. Rowe Forest Regions

FOREST REGION	DESCRIPTOR AND KEY SPECIES	TOTAL AREA (MILLION HECTARES)	PERCENT OF CANADA'S LAND AREA
Boreal (total)	Predominantly coniferous, the boreal forest extends across Canada from Newfoundland to the Rocky Mountains and from the southern grasslands to the tundra. Characteristic species include white and black spruce, tamarack, balsam fir, and jack pine, with a mixture of broad-leaf trees such as aspen and poplar. This forest region is divided into three sub-regions.	529	53
Boreal (Boreal – Predominantly Forested)	Predominantly forested zone in southern half of the boreal region.	(289)	(29)
Taiga (Boreal – Forest and Barren)	A zone of open forest, wetland, and barren land in the northern half of the boreal region.	(218)	(22)
Aspen Parkland (Boreal – Forest and Grassland)	A mixed grassland and open forest marking the transition to the true grassland ecosystems of southern Alberta, Saskatchewan, and Manitoba.	(22)	(2.2)
Great Lakes (Great Lakes-St. Lawrence)	Mixed coniferous and deciduous forests found in central Ontario and along the St. Lawrence River valley in Quebec. Characteristic species include eastern white and red pines, eastern hemlock and yellow birch. Associated broad-leaf species include maple, oak, basswood, aspen, ash, and elm.	46.6	4.7
Subalpine	Coniferous forest found on the mountain slopes of Alberta and British Columbia. Characteristic species are Engelmann spruce, alpine fir, and lodgepole pine.	25.1	2.5
Columbian	In the wet interior forests of British Columbia are coniferous forests with characteristic species consisting of western red cedar, western hemlock, and interior Douglas fir.	5.5	0.6
Montane	Largely restricted to the dry, central plateau of British Columbia and a few southern mountain valleys adjacent to the Alberta border. Douglas fir is a characteristic tree species, as well as lodgepole pine, Engelmann spruce, and alpine fir. In the southern portion of the region, ponderosa pine is common.	15	1.5
Acadian	Encompasses most of the maritime provinces and is closely related to the Great Lakes Forest Region. It has a mixture of conifers and deciduous trees. Red spruce is a characteristic species; associated with it are balsam fir, yellow birch, sugar maple, with some red pine, eastern white pine, and eastern hemlock.	12.2	1.2
Coast	Found along the mainland coast of British Columbia, this forest region receives more than 2 meters of annual precipitation and is renowned for its old and large trees and diverse fauna. Characteristic tree species include western hemlock, western red cedar, Douglas fir, and Sitka spruce.	12.6	1.3
Carolinian (Deciduous)	Confined to southwestern Ontario, this region contains deciduous trees common to the Great Lakes Forest Region. Other species, such as the tulip tree and cucumber tree, have their northern limits here, but are more common in the United States.	5.1	0.5

Sources: J.S. Rowe, *Forest Regions of Canada* (Ottawa: Canadian Forestry Service, Dept. of Fisheries and Environment, 1977).
Canadian Council of Forest Ministers, *Compendium of Canadian Forestry Statistics 1996* (Ottawa: Natural Resources Canada, 1997).

BOX 2 Is Climate Change a Threat to Forests?

Global climate change is expected to have a major impact on Canada's forests. Average global temperature could increase 1°C to 3.5°C by the end of the next century, but warming will not be uniform around the globe or within Canada.[1] High latitudes are expected to warm much more than the global average. Canada's mean annual temperature may increase 5°C to 10°C over the next century, more than three times the global average.[2]

Given Canada's size and diversity of landscapes and ecosystems, climate change is expected to affect every region differently. Climate change could lead to greater stress because of droughts and more frequent and severe fire and insect disturbances. In some areas, climate change could result in increased vegetative growth rates. Some models also predict more frequent, extreme storms and wind damage, especially in coastal areas, but there is no general consensus in the modeling community on this point.

Fire is a natural part of most forest ecosystems in Canada, especially in the boreal forests, where lightning fires that burn over large areas are natural events. If climate change occurs as expected, the scale and intensity of fires is expected to increase. Most ecosystem or fire-modeling scenarios show increases in primary productivity, leading to greater accumulations of fuel. In drought years, more dry fuel means greater risk of fire. These changes in disturbance regimes will likely have a larger impact on Canadian forests than climate change itself.[3]

For the boreal region west of Lake Superior, the frequency and intensity of fires is expected to increase, resulting in a longer fire season, more fires, and more area burned each year.[4] Given predictions of a doubling in CO_2 concentrations in the next century, some scientists predict close to a 50 percent increase in fire intensity, with a possible similar increase in area burned.[5] On the southern boundary of the boreal forest, the severity of fires may double.[6]

The pace of change in Canada's boreal forests may occur faster than forests can adapt.[7] On the southern fringe, northern deciduous species and balsam fir may displace the current mix of spruce, pine, larch, poplar, and birch. In mid-continental areas, grasslands may replace the southern boreal forest. And in the eastern boreal regions, the tree line may advance into areas previously occupied by tundra. Boreal tree species would have to migrate northward at between 1.5 and 5.5 kilometers/year over the next 50 to 100 years to remain within similar climate conditions, which is most likely beyond the adaptive capabilities of most tree species.[8] The net result is a boreal forest that may shrink to half its current extent.[9]

[1] Science Branch, Canadian Forest Service, *Climate Change and Forests: Context for the Canadian Forest Service's Science Program* (Ottawa: Natural Resources Canada, 1999), p. 4.

[2] Science Branch, Canadian Forest Service, *Climate Change and Forests: Context for the Canadian Forest Service's Science Program* (Ottawa: Natural Resources Canada, 1999), p. 4.

[3] M.G. Weber and M.D. Flannigan. 1997. "Canadian boreal forest ecosystem structure and function in a changing climate: impacts on fire regimes." *Environmental Review* 5: 145-166.

[4] Science Branch, Canadian Forest Service, *Climate Change and Forests: Context for the Canadian Forest Service's Science Program* (Ottawa: Natural Resources Canada, 1999), p. 6.

[5] M.D. Flannigan and C.E.Van Wagner. 1991. "Climate change and wildfire in Canada." *Canadian Journal of Forest Research* 21: 66-72.

[6] M.A. Fosber, B.J. Stocks, and T.J. Lynham., "Climate change-fire interactions at the global scale: predictions and limitations of methods," in *Fire in the Environment*, Crutzen, P.J., J.G. Goldammer, eds., (Chichester: John Wiley and Sons, 1993), pp. 123-137.

[7] Canadian Forest Service, *The State of Canada's Forests: The People's Forests 1997-1998* (Ottawa: Natural Resources Canada, 1998), p. 86.

[8] G. Esser, 1992. "Implications of climate change for production and decomposition in grasslands and coniferous forests." *Ecological Applications* 2: 47-54.

[9] B. Rizzo and W. Wiken. 1992. "Assessing the sensitivity of Canada's ecosystems to climate change." *Climate Change* 21: 37-55.

Technical Note 1. Forest Regions/Types

Rowe Forest Regions are major geographic zones with similar types of dominant tree species. They describe climax forest types, and as such are a measure of potential rather than actual forest cover. Rowe's original description included 12 forest regions. Of these 12, 8 regions are now generally accepted as forest regions. In this report, we present data in terms of 10 forest types; the boreal regions are treated as separate categories. For the sake of simplicity, we use popular names in some cases rather than the original terms. We have used Rowe's data to characterize current forest cover (derived from the 1995 *Land Cover of Canada* map[1]) by forest type. Several other forest classification schemes are also widely used in Canada.

[1] J. Cihlar and J. Beaubien *Land Cover of Canada Version 1.1* (Ottawa: Canada Centre for Remote Sensing, 1998).

FOREST CONDITION

Although often viewed as an endless expanse of wilderness, Canada's forests are undergoing development of their commercially valuable resources. This development now extends into some of the northernmost and most ecologically sensitive forests in the country. Development activities include logging (primarily in the southern third of the country), mining, oil and gas exploration, hydroelectric development, and conversion of forests to agriculture and other land use (primarily in the south). Other activities include hunting, trapping, fishing, and tourism.

In this section, we present data on the magnitude of forest development. Due to data limitations, we focus on readily mapped activities, such as roads, mines, and tenures (referring to forests with area- or volume-based licenses that are generally managed for wood production). We have also included settlements (from small communities to larger cities) as a proxy measure of nonindustrial uses (recreation, hunting, and other small-scale uses). Appendix 3 provides an overview of the positive and negative impacts of the major industries and activities considered here.

Indicator 3: Cumulative Forest Development

By compiling a series of relevant, readily available national and provincial datasets, we derived a picture of cumulative forest development in Canada. *(See Technical Note 2.)*

More than 60 percent of Canada's forests are either tenured or within 10 kilometers of a development activity

Sixty-two percent of Canada's forest area shows some mark of human activity—falling within a tenure or within 10 kilometers of a road, other access route, settlement, mine, or hydroelectric dam. *(See Figure 2.)* Undeveloped areas are almost exclusively confined to the northern Boreal and Taiga Forest Regions.

Just over half of all forest area is within 10 kilometers of a development activity

Some 51 percent of forest area is situated within 10 kilometers of a development activity. The total is likely an underestimate because we lacked complete data, particularly for access routes. *(See Technical Note 6.)* Roads and other access routes account for the greatest share of this development, followed by mines, settlements, and hydroelectric development. All told, there are at least 196 active mines, 1,100 hydroelectric dams, and 4,045 settlements in areas that are currently, or were historically, primarily forested (10 Forest Regions).

The Great Lakes Forest Region has the most development, with at least 38 active mines,

625 hydroelectric dams, and 793 settlements. *(See Table 2.)* Boreal forests have the least development. Even in the more remote forest areas of the Boreal and Taiga Forest Regions, however, there are over 80 active mines, over 300 hydroelectric dams, and over 1,400 settlements.

Approximately 52 percent of forest area is tenured for wood production

About 52 percent of forest area is under some form of tenure. Tenures include areas of land that have not been and are not likely to be logged because they are not sufficiently productive. *(See Section 3 for further information.)*

Table 2. Developments in Forest Regions

FOREST REGION	ACTIVE MINES (1997)	INACTIVE MINES	HYDROELECTRIC DEVELOPMENT	SETTLEMENTS
Boreal	82	737	257	1,376
Aspen Parkland	13	27	7	731
Taiga	7	129	58	97
Subalpine	4	352	14	71
Montane	12	278	21	220
Coast	3	325	22	237
Columbian	8	598	7	117
Carolinian	7	160	42	105
Great Lakes	38	2,633	625	793
Acadian	22	237	47	298
Total	**196**	**5,476**	**1,100**	**4,045**

Source: GFW Canada

Technical Note 2. Cumulative Development

We compiled a range of existing datasets on mines, hydroelectric development (dams and generating stations), settlements, access, and tenures. We used these datasets to calculate the amount of existing forest that is under some form of development. Forest is defined as classes 1-12 of the 1995 *Land Cover of Canada* map. For each dataset, we classified as "developed" any 10-kilometer by 10-kilometer forest area that included a development activity. For example, in the mine analysis, any 100 km^2 cell that included an active or inactive mine was considered "developed." The various layers were then overlaid to produce a total percentage of area under development.

The tenure layer is a combination of existing area- and volume-based tenures as well as other areas managed for timber production. This is what we call the commercial forest zone. Our analysis eliminated the nonforest portions of the commercial zone area. As noted in the text, tenures do not necessarily imply that all forest area will be harvested. Other types of development are likely underestimated. For example, many hydroelectric dams impact an area far larger than 100 km^2.

Figure 2. Percentage of Forest Area Affected by Development

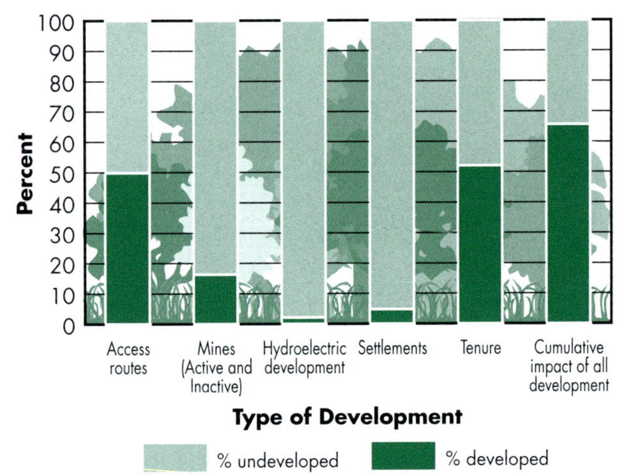

Source: GFW Canada

Indicator 4: Watershed Development

Forested watersheds provide a range of important ecosystem services to humans, including protection of downstream areas from erosion, maintenance of water flow, and protection of habitat for salmon and other valuable aquatic species. The absence of development provides one crude indicator of the degree to which these services are maximized. The presence of development is an indicator of impacts such as clearing and resource extraction.

The potential impacts of development—in terms of erosion, pollution, and habitat loss for aquatic and other species—are difficult to quantify nationally. The extent and severity of impacts depend heavily on land use planning, resource management practices, and the differing types of development.

The distribution of industrial activities within Canada's major watersheds provides a further indication of the portion of Canada's forest ecosystems affected by human activities. Using various datasets, we analyzed the distribution of these activities in large watersheds within the 10 Forest Regions. *(See Technical Note 3.)* Map 2 illustrates the development status of large forested watersheds based on this analysis.

Some human development is evident in 95 percent of forested watersheds

In 42 percent of forested watersheds, development is occurring in over half the area of the watershed.

In 555 of 820 watersheds, at least 10 percent of the total area includes settlements, tenures, energy, or mineral development. In a majority of watersheds (56 percent), development occurs in at least a fourth of the watershed area. *(See Table 3.)*

Most undeveloped watersheds are at the northern forest edge (Taiga Forest Region)

Not surprisingly, the majority of the undeveloped watersheds are located within relatively remote forests in the north, predominantly in the Taiga Forest Region.

Technical Note 3. Watershed Development

GFW Canada examined the distribution and effects of activities on large (tertiary) watersheds with an average size of 8,400 km^2. Only forested watersheds were included in the analysis; we excluded tundra and grasslands. These watersheds total approximately 7.1 million km^2, or three fourths of Canada's land area. Rather than attempting to set thresholds, we have simply assigned watersheds to five classes of development: 0 percent, 0-10 percent, 10-25 percent, 25-50 percent, and 50 percent and greater. Our cumulative development analysis was used as the basis for classifying watersheds into these categories. See *Appendix 1* for more detail.

Table 3. Status of Forested Watersheds

PERCENTAGE OF WATERSHED DEVELOPED	NUMBER OF WATERSHEDS IN CATEGORY	PERCENTAGE OF TOTAL
0	43	5
0 - 10	222	27
10 - 25	97	12
25 - 50	117	14
50 and greater	341	42

Source: GFW Canada

Indicator 5: Cleared Areas and
Indicator 6: Accessed Forest

The effects of development activities on biodiversity, wilderness values, and a range of other ecosystem services can be measured in terms of forest loss (conversion to nonforest land use) and fragmentation by roads and other access routes. Forest conversion results in outright loss of habitat for forest-dependent species, posing a far greater threat to biodiversity than fragmentation alone. Fragmentation—by roads, railroads, pipelines, and powerlines—can impact species in a number of ways:

- Habitat degradation associated with access. Wolves, for example, are almost never found where there are more than 0.45 km of roads per km^2 of area.[20]
- Creating barriers to species that avoid areas where humans are present. Some wildlife species, such as grizzly bears, are highly sensitive to roads.
- Increased illegal hunting pressure facilitated by roads and other access routes. Moose, wolves, caribou, mountain goats, and bighorn sheep are particularly vulnerable to this kind of pressure.[21]

Extensive clearing and/or fragmentation also free carbon locked in vegetation and soils, and as such represent a loss of carbon sequestration services offered by forests. *(See Box 3.)* Finally, areas that are distant from clearing and roads have a high wilderness value.

Using the 1995 *Land Cover of Canada* map and compiled datasets of access routes, we analyzed the percentage of Canada's forests that have been converted and fragmented. *(See Technical Note 4.)*

Map 3 presents our analysis of the extent of converted and accessed forests in Canada.

Technical Note 4. Conversion and Access

GFW Canada undertook an assessment of the extent of converted and accessed forests and remaining unaccessed forests across Canada. Converted forest results are based on an overlay of urban and agricultural classes from the 1995 *Land Cover of Canada* map with Rowe's forest regions. This map provides information on actual forest and other land cover within Canada. This map is derived from satellite imagery and includes 31 classes of land cover. Of these, GFW considered 13 forest and another 3 woodland/cropland mosaics.

Linear features such as roads, pipelines, railways, and hydropower lines were mapped to identify accessed forests. Because most human influences occur close to roads and decline rapidly with distance, 1 kilometer was assumed to be a critical distance beyond which human influences are minimal.[1] Therefore, GFW Canada classified any 1 kilometer x 1 kilometer area that included known access features as "accessed."

[1] United States Department of Agriculture-Forest Service, *Forest Service Roads: A Synthesis of Scientific Information* (Washington, DC: USDA-Forest Service, 1999). Online at: http://www.fs.fed.us/news/roads. (January 14, 2000). See also: Silva Forest Foundation, *Assessing the Ecological Impacts of Timber Management: Apparent Impacts, Actual Impacts and Precautionary Forest Development Planning.* (Winlaw, BC: Silva Forest Foundation, 1999)

Indicator 5: Cleared Areas

At least 6 percent of Canada's forest area has been cleared for farmland and settlements

Over 26 million hectares of historically forested lands are now classified as cultivated and built-up area. *(See Table 4.)* This figure does not include extensive areas that remain as a mix of forest, farmland, and suburban land, or forest converted to grassland. Compared to most countries, however, a relatively small portion of Canada's forest area has been cleared to date, primarily because of low population pressure and the fact that most of this land is unsuitable for farming. *(See Technical Note 5.)*

Canada's most species-rich forests are also the most modified

Northern forests have undergone very little clearing

The Carolinian and Aspen Parkland Forest Regions have lost 55 percent or more of their forest cover to clearing for agriculture and settlements. Less than 3 percent of the Boreal Forest Region has been cleared to date. The Taiga Forest Region has had very little clearing.

Three provinces have had over 19 percent of their forests converted

Six provinces have had 5 percent or less of their forests converted

Prince Edward Island, Alberta, and Saskatchewan have lost the largest percentage of forest to conversion. *(See Table 5.)* British Columbia, Newfoundland, New Brunswick, Nova Scotia, Ontario, and Quebec have each lost less than 5 percent through conversion.

BOX 3 The Role of Canada's Forests in Mitigating Climate Change

The total store of carbon in Canada's forests, soils, peatlands, and forest products in the 1990s is estimated to be about 221 billion tons, including about 135 billion tons stored in peatlands.[1] The boreal forests of Canada and Russia contain nearly 40 percent of the planet's terrestrial carbon stocks.[2] Peatbogs are an especially important carbon sink in the boreal and taiga regions. The energy content of Canada's peatlands is estimated to be equivalent to 60 percent of the world's hydroelectric capacity.[3]

Whether Canada's forests are a net source or sink of carbon—that is, whether a carbon pool is considered to be gaining carbon (a "sink") or losing carbon (a "source")—is currently unclear. In October 1999, the Canadian government submitted greenhouse gas inventory data to the United Nations Framework Convention on Climate Change showing that the forestry and land use sector was a net carbon sink. The government reported a net removal of CO_2 from the land use change and forestry sector over the period from 1990 to 1997. However, the trend in net removals was down by some 56 percent, from 44,000 gigagrams in 1990 to 19,000 gigagrams in 1997.[4]

Other modeling of carbon budgets indicates that Canada's forests were a net carbon sink from 1920 to the mid-1980s, absorbing 118 million tons annually.[5] Starting in the mid-1970s, the rate of carbon uptake declined. From the mid-1980s onward, Canada's forests have become a carbon source, releasing 45 million tons of carbon per year.[6] This flux of carbon is the result of unusually large-scale forest disturbances by fire and insects since 1970 and increased drying of peatlands.[7] Canada's forests, however, could once again become a forest sink if peatlands experience less drying and fire and insect disturbances decline.

Recent research indicates that a large increase in fire and insect disturbances has caused the boreal and subarctic forests to become a net source of atmospheric carbon.[8] Harvesting appears to have played a minor role in the change from carbon sink to source in this region.

[1] Canadian Council of Forest Ministers, *Criteria and Indicators of Sustainable Forest Management in Canada* (Ottawa: Natural Resources Canada, 1996), p. 60.

[2] E.S. Kasischke, N.L. Christensen, and B.J. Stocks. 1995. "Fire, global warming, and the carbon balance of boreal forests." *Ecological Applications* 5(2): 437-451.

[3] P. Kauppi, and M. Posch. 1989. "Boreal forests and the Global Carbon Cycle." *Science*: 243(4898): 1535-1536.

[4] United Nations Framework Convention on Climate Change (UNFCCC), *National Communications from Parties Included in Annex I to the Convention: Greenhouse Gas Inventory Data, 1990-1997* (New York: UNFCCC, 1999) Table B.7.

[5] W.A. Kurz and M.J. Apps. 1995. "An analysis of future carbon budgets of Canadian boreal forests." *Water Air Soil Pollution* 8(2): 321-332.

[6] Science Branch, Canadian Forest Service, *Climate Change and Forests: Context for the Canadian Forest Service's Science Program* (Ottawa: Natural Resources Canada, 1999), p. 8.

[7] W.A. Kurz and M.J. Apps. 1995. "An analysis of future carbon budgets of Canadian boreal forests." *Water Air Soil Pollution* 8(2): 321-332.

[8] Science Branch, Canadian Forest Service. *Climate Change and Forests: Context for the Canadian Forest Service's Science Program* (Ottawa: Natural Resources Canada, 1999), p. 7.

Technical Note 5. Cleared Forest Area

Data underlying a 1997 WRI report compiled with the help of the World Conservation Monitoring Centre indicate Canada has lost just under 10 percent of its forest cover. GFW Canada's analysis provides a much more conservative estimate of around 6 percent. Our figure is more accurate and an underestimate, since we did not factor in forest loss in areas naturally dominated by nonforest or conversion of forest to grassland or forest/cropland mix.

Between 54,600 and 80,500 hectares of forest are cleared annually

Canadian government reports state that 88,000 to 103,000 hectares of forest are converted annually to agricultural and other land use.[22] A more recent study estimates that between 54,600 to 80,500 hectares are cleared per year.[23] This is a relatively small fraction (0.01 to 0.02 percent) of Canada's forestland.

Indicator 6: Accessed Forest

Within seven forest regions, over half of the forested area is fragmented by access routes

Northern forests remain largely unfragmented by access routes

Canada's most productive forests are the most heavily fragmented. In the Carolinian and Aspen Parkland Forest Regions, access routes fragment over 95 percent of remaining forests. Most of the remaining forests of the Coast (54 percent), Acadian (75 percent), and Columbian (82 percent) Forest Regions have been fragmented. *(See Table 4.)* These forests exhibit some of the highest species diversity in Canada. Forests of the Coast Forest Region, as noted earlier, harbor large tracts of globally important temperate rainforest. Only the Boreal Forest Region (31 percent) and the Taiga Forest Region (1 percent) remain largely unaccessed.

Five provinces have more than 60 percent of their remaining forests accessed

The forests of New Brunswick and Alberta are both over 80 percent accessed. British Columbia, Nova Scotia, and Prince Edward Island all have over 60 percent of their remaining forests fragmented by access routes. *(See Table 5.)* Newfoundland and the three territories are the least accessed. *(See Technical Note 6.)*

Box 4 provides a detailed analysis of northwestern Alberta and northeastern British Columbia, where oil- and gas-related activities have had significant impacts on forests.

Technical Note 6. Caveat on Access Figures

We overestimate fragmentation in northern Alberta and northeastern British Columbia. In this area, our data layers include oil and gas seismic exploration lines. We underestimate access in other areas. Some of our datasets for access are quite coarse, such as the Digital Chart of the World, which we used for Newfoundland. Quebec access is also underestimated, since we produced our access estimates for this province using satellite imagery. (*See Appendix 1 for further details.)*

Table 4. Forest Conversion and Access by Forest Region

ROWE FOREST REGION	CONVERTED LAND (000 HECTARES)	PERCENT CONVERTED	TOTAL REMAINING FOREST (000 HECTARES)	ACCESSED FOREST (000 HECTARES)	PERCENT ACCESSED FOREST
Boreal	7,711	3	200,133	61,423	31
Aspen Parkland	12,282	62	7,517	7,332	98
Taiga	1	0	115,067	1,333	1
Subalpine	95	0	13,791	6,678	48
Montane	309	1	11,856	9,216	78
Coast	186	3	7,034	3,817	54
Columbian	32	3	3,688	3,040	82
Carolinian	1,980	57	764	743	97
Great Lakes	3,299	9	32,997	18,127	55
Acadian	612	5	10,691	8,045	75
Total	**26,507**	**6**	**403,538**	**119,754**	**30**

Source: GFW Canada

Table 5. Forest Conversion and Access by Province

PROVINCE	CONVERTED LAND (000 HECTARES)	PERCENT CONVERTED	TOTAL REMAINING FOREST (000 HECTARES)	ACCESSED FOREST (000 HECTARES)	PERCENT ACCESSED FOREST
Northwest Territories	1	0	46,517	1,217	3
Yukon Territory	1	0	12,929	1,252	10
British Columbia	849	2	48,184	30,137	63
Quebec	1,468	2	80,948	16,898	21
Newfoundland	15	0	16,813	668	4
Ontario	3,858	5	78,721	21,849	28
Nova Scotia	203	4	4,879	3,035	62
New Brunswick	239	3	6,612	5,797	88
Prince Edward Island	211	46	250	153	61
Alberta	9,035	21	33,259	27,518	83
Saskatchewan	7,270	19	31,558	5,841	19
Manitoba	3,304	8	40,289	5,074	13
Nunavut	0	0	1,977	0	0
Total	**26,454**	**6**	**402,936**	**119,439**	**30**

Source: GFW Canada
Note: The slight variations in totals between Tables 4 and 5 are due to the coarseness of provincial borders.

BOX 4 Fragmenting the Forest of the Western Canadian Sedimentary Basin

The Western Canadian Sedimentary Basin occupies 545,000 km^2 of forest in Alberta and northeastern British Columbia.[1] The basin is Canada's major oil- and gas-producing region, as well as having forestry and mining activities. Increased industrial activities in the region "are unprecedented in terms of both their huge scale and rapidity of development,"[2] primarily to satisfy market demand in the United States.[3]

Fifty years ago, the forests of Alberta and northeastern British Columbia were still mostly free of roads and other linear access features. Today, the area is quite changed, as noted in the aerial photography sequence of Alberta's Swan Hills from 1949 to 1991. *(See Figure 3.)* This sequence provides a classic example of habitat loss and fragmentation resulting from industrial development. While dramatic, the example illustrated by this sequence is by no means unique within the basin's forest.[4]

One indicator of industrial activity is what ecologists call "linear disturbances" or access densities. These disturbances include roads, seismic lines, railways, pipelines, and powerlines. Linear features fragment the forest and create human access to formally secluded areas. Using data from over 650 map sheets, GFW analysis found that most of the forests in the basin have been fragmented by more than 1.3 million kilometers of linear disturbances. The resulting average access density in these forests is currently 2.4 km/km^2. *(See Map 4.)* This industrial access has fragmented the basin's forests into patches. Only 745 of the 11,978 remaining patches, or 6 percent, are greater than 9 km^2 in size. These patches have a mean size of 64 km^2.

[1] This figure does not include the 4,159 km^2 that represents the area of the major lakes found within the study area.

[2] Alberta Environmental Protection, *The Boreal Forest Natural Region of Alberta* (Edmonton: Alberta Environmental Protection, 1998), p. 1.

[3] National Energy Board, *National Energy Board Annual Report* (Calgary: National Energy Board, 1998), pp. 12, 16. Online at: http://www.neb.gc.ca/about/ar/1998/index.htm (January 23, 2000).

[4] Alberta Environmental Protection, *Prospects for Protection: The Foothills Natural Region of Alberta* (Edmonton: Alberta Environmental Protection, 1996), pp. 63-67.

Figure 3. Photo-mosaic Showing Increasing Conversion and Fragmentation in Alberta

From 1949 to 1991, Alberta's Swan Hills changed from a roadless wilderness to an intensely fragmented landscape. By 1964, activities included oil and gas exploration, well sites, and roads. By 1982 and then 1991, clearcuts from logging and more roads were visible. The photo-mosaic shown here is at Imperial Tower, 35 kilometers north of Whitecourt, Alberta. White stars in the photographs indicate the same reference point of 54° 27' N, 11° 15' 36". Photographs are not all the same scale.

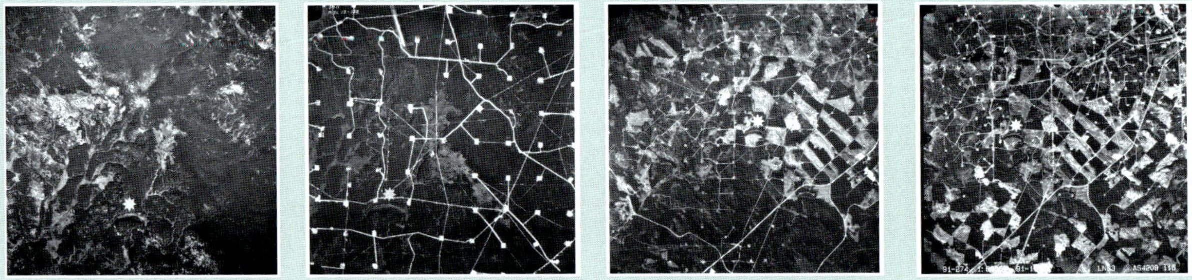

1949 *1964* *1982* *1991*

Sources: Alberta Environmental Protection, *Prospects for Protection: The Foothills Natural Region of Alberta* (Edmonton: Alberta Environmental Protection, 1996), pp. 63-67.

PHOTOGRAPHS: COMPILED BY RICHARD THOMAS

Indicator 7: Unfragmented Forest

GFW Canada also undertook an analysis of the size of remaining contiguous blocks of forest cover by forest region. *(See Appendix 1.)* Large areas of unfragmented forest are important for the biodiversity values they provide—both because they potentially serve as habitat for viable populations of their species and because they can serve as representative samples of relatively undisturbed forest ecosystems.

Map 5 depicts the distribution of remaining unfragmented forest in three size categories: 200-500 km^2, 500-10,000 km^2, and over 10,000 km^2. These categories correspond approximately to the minimum size contiguous blocks of forest must be to maintain natural processes such as fire, wind, and insect infestations. The magnitude and nature of natural disturbances varies by forest type. For example, in the boreal forest, large fires are not uncommon, and large contiguous forest areas (10,000 km^2 of forest) are necessary to allow large fires to continue their natural role in maintaining habitat for the full range of native biodiversity. According to expert estimates in the literature, this minimum size is probably closer to 500 km^2 for coastal temperate rainforests and 200 km^2 for eastern forests.[24]

The results from the area size analysis indicate the proportion of remaining forests that can most likely support a full range of species and full natural renewal processes. The reverse, however, is not necessarily true. These areas do not indicate the proportion of forests that do not meet these criteria. Access routes do not necessarily create a barrier to forest species and smaller tracts of fragmented forest may maintain much or all of their biodiversity with improved management.

Over 60 percent of Canada's forests are in very large (10,000 km^2) unfragmented blocks

All told, almost 1.9 million km^2 of forest are in very large (10,000 km^2 or more) blocks unfragmented by access routes. However, almost all of this area is found within the Taiga Forest Region and the northern part of the Boreal Forest Region. *(See Table 6.)*

About one seventh of Canada's forest area is highly fragmented by roads and access routes—in areas under 200 km^2 in size

The Carolinian Forest Region and the majority of the forest in the Columbian and Acadian Forest Regions are dominated by fragmented forest under 200 km^2 in size.

One third of the Coast Forest Region is in unfragmented blocks exceeding 500 km^2

Almost half is in areas under 200 km^2 in size.

The Coast Forest Region—the most species-rich in Canada—is relatively fragmented. Of the total estimated original forest area, only 15 percent remains in unfragmented areas over 200 km^2. Almost half of remaining forest is in blocks under 200 km^2 in size, and 53 percent is in areas greater than 200 km^2.

Indicator 8: Forest-Dwelling Species at Risk

Estimates of forest fragmentation and modification provide indirect indicators of the potential threats of development to biodiversity. With careful management, these threats may not translate to impacts. Data on numbers of species at risk provide a direct measure of development impacts on biodiversity. This measure only applies to species and not to ecosystems. No global standard exists for assessing where, or to what degree, habitats are threatened.

Canada's forests provide habitat for about two thirds of Canada's estimated 140,000 species of plants, animals, and microorganisms, only half of which are classified.[25] Species deemed to be at risk are assessed by the national Committee on the Status of Endangered Wildlife in Canada (COSEWIC) and added to a national list on a yearly basis.[26] COSEWIC assigns each species to a category: "vulnerable" species are of special concern due to their sensitivity to human activities or natural events; "threatened" species are likely to become endangered if limiting factors are not reversed; and "endangered species" face imminent extinction or extirpation. (Extirpation refers to local extinction.)

Table 6. Unfragmented Forest by Forest Region (km²)

FOREST REGION	1-200 km²	200-500 km²	500-10,000 km²	OVER 10,000 km²
Boreal	146,362	54,362	309,140	877,245
Aspen Parkland	1,680	0	0	0
Taiga	86,598	21,468	68,792	960,486
Subalpine	23,532	6,118	34,963	0
Montane	10,114	3,463	12,819	0
Coast	15,136	5,111	11,924	0
Columbian	5,025	445	1,014	0
Carolinian	204	0	0	0
Great Lakes	53,887	22,554	67,166	0
Acadian	21,590	2,216	2,649	0
Total	**364,128**	**115,737**	**508,467**	**1,837,731**

Source: GFW Canada

Less than 1 percent of species are classified "at risk"

25 percent of all species at risk are forest dependent

As of April 1999, there were 339 species on the national list of species at risk in Canada.[27] The list includes 84 forest-dwelling species at risk. *(See Table 7.)* Of the 32 new species added to the list in 1999, 7 are dependent on forests.[28] These figures underestimate threats to forest biodiversity for two reasons:

1) The status of species is not systematically assessed, in part for lack of data; and

2) There is generally a lag time between when human pressures begin taking a toll on a species and when those effects are first noticed through population monitoring.

Most species at risk are found within two biodiversity-rich habitats, which have undergone high rates of clearing and fragmentation

The two Forest Regions that contain the most species in the at-risk category are the temperate rainforests of the Coast Forest Region in British Columbia and the Carolinian Forest Region of southwestern Ontario.[29] Roughly 60 percent of Canada's endangered forest-dwelling species are found in the Carolinian Forest Region.[30]

Within the Coast Forest Region (coastal temperate rainforest), there are 56 endangered, threatened, or extirpated species, and 150 vulnerable species.[31] The majority of these are vascular plants. In addition, there are over 150 plant communities at risk within the coastal temperate rainforest zone.

Two flagship wilderness species are threatened throughout much of Canada

The two most widespread forest species at risk are the woodland caribou and wolverine.[32] Their threatened status reflects the changing conditions of forests as roads, logging, mining, and other developments fragment the forest. The woodland caribou requires mature or old-growth coniferous forests, which are disappearing across much of their range in Quebec, Ontario, Saskatchewan, Alberta, and British Columbia.[33] Wolverines use a variety of habitats, but like some other species such as grizzly bear they are sensitive to human disturbance. The abundance and distribution of these animals decline even with low levels of disturbance and forest fragmentation.

In addition to species at risk, 52 animal species and 12 plant species occupy a small portion of their former range.[34] Their ranges have been reduced primarily by development activities that have either eliminated or degraded their forest habitats.

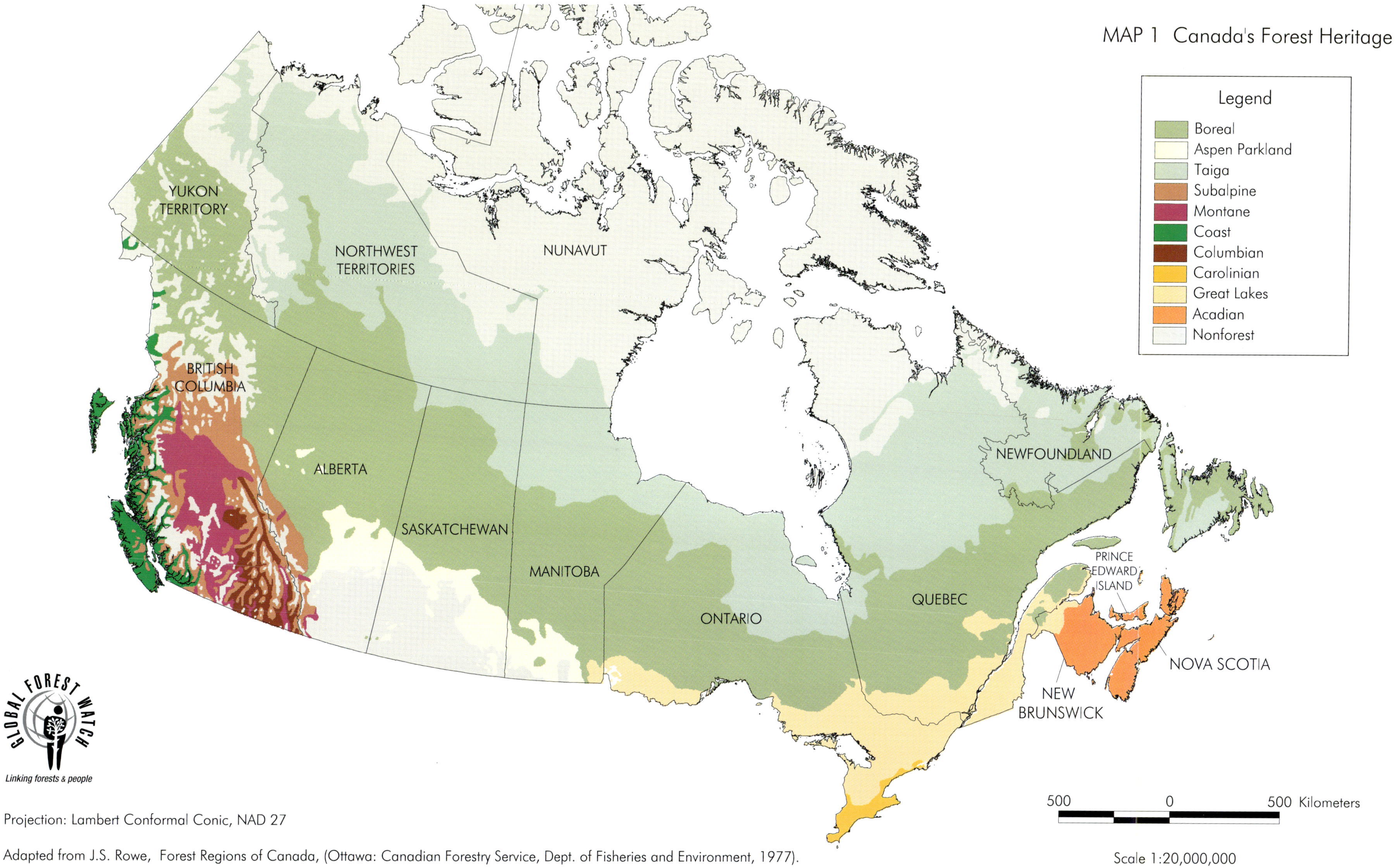

Projection: Lambert Conformal Conic, NAD 27

Adapted from J.S. Rowe, Forest Regions of Canada, (Ottawa: Canadian Forestry Service, Dept. of Fisheries and Environment, 1977).

MAP 2 Development Status of Forested Watersheds

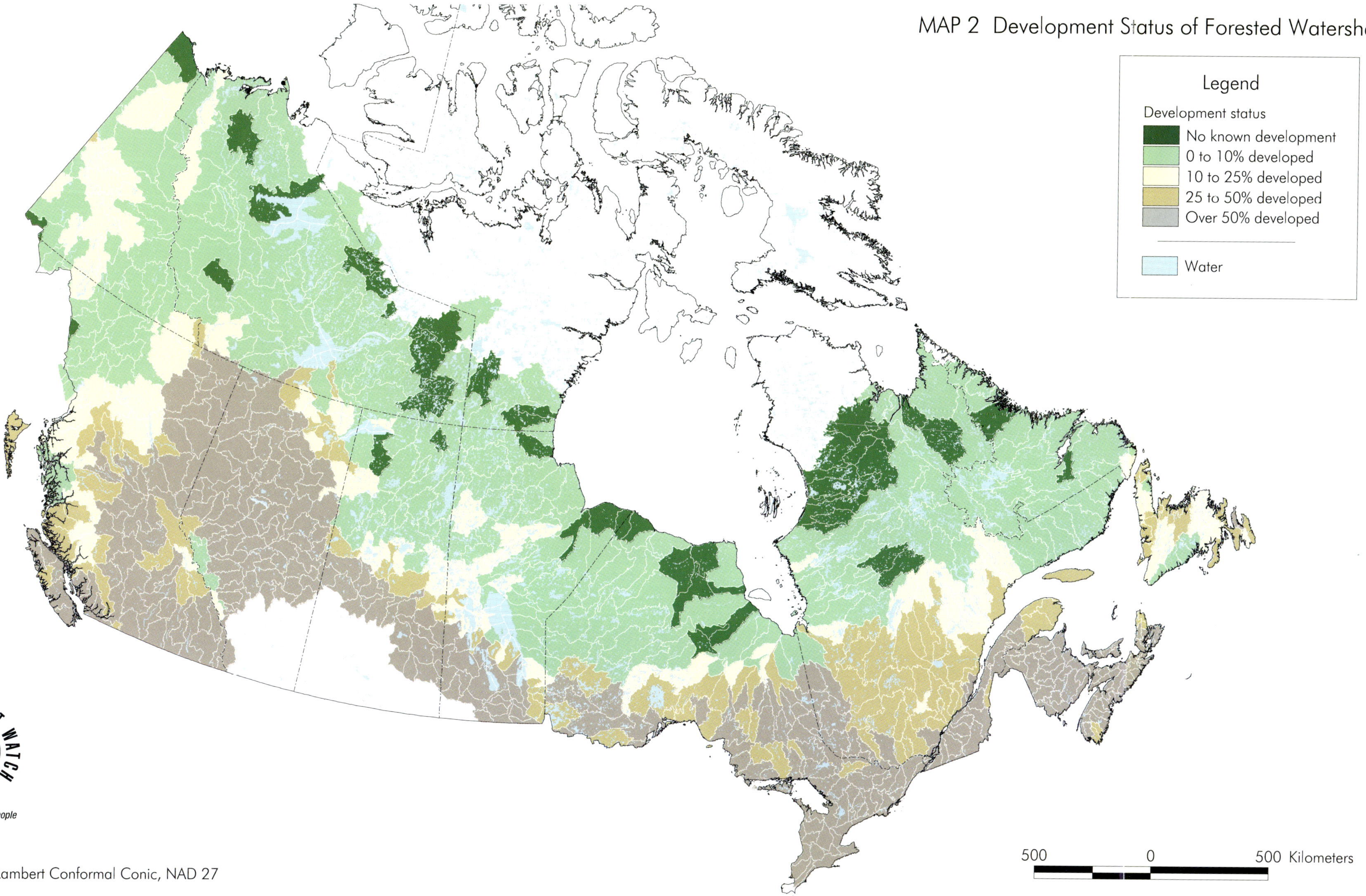

GLOBAL FOREST WATCH
Linking forests & people

Projection: Lambert Conformal Conic, NAD 27

Produced by Global Forest Watch Canada

MAP 3 Converted and Accessed Forests

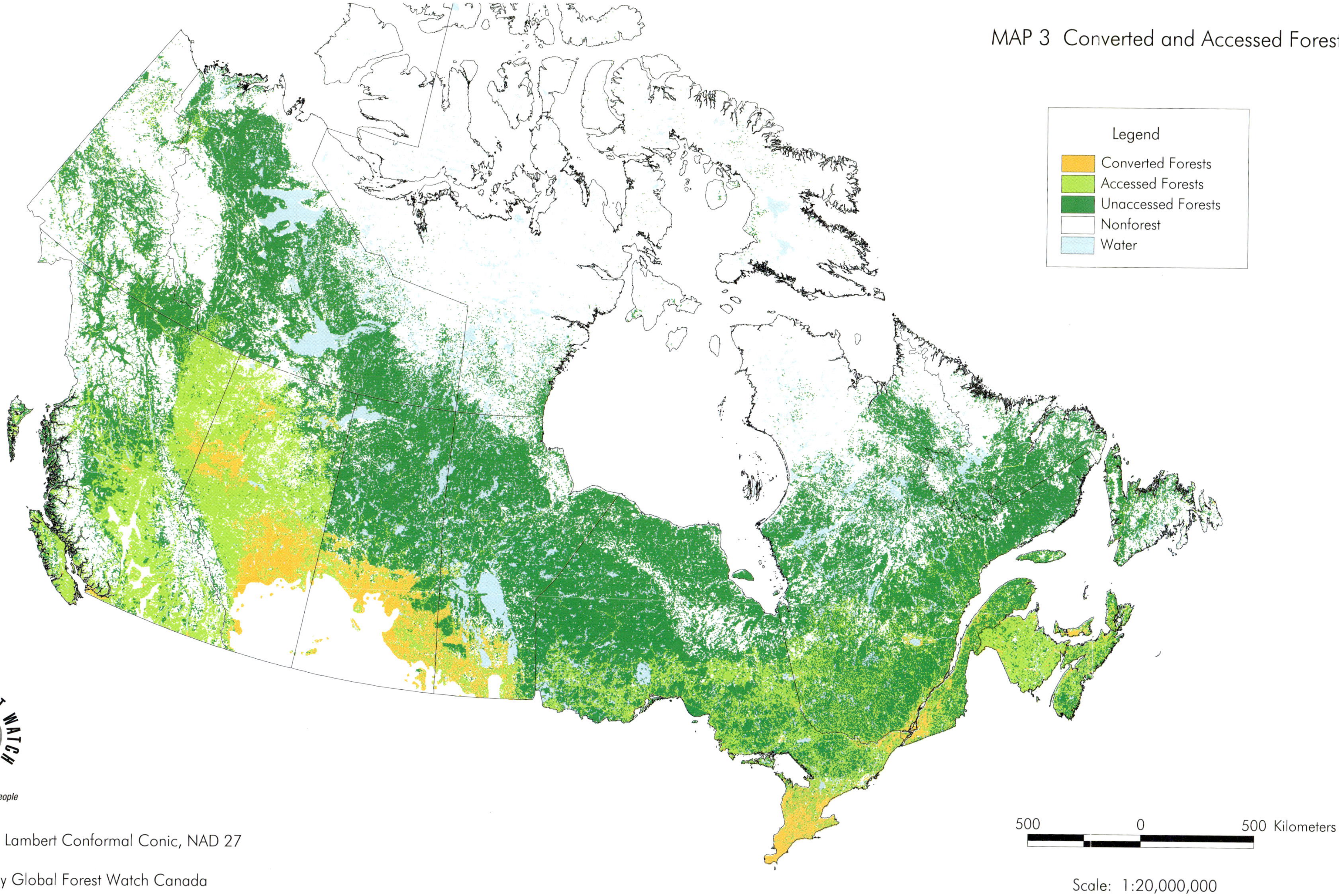

GLOBAL FOREST WATCH
Linking forests & people

Projection: Lambert Conformal Conic, NAD 27

Produced by Global Forest Watch Canada

MAP 4 Access Densities in the Western Canadian Sedimentary Basin (WCSB)

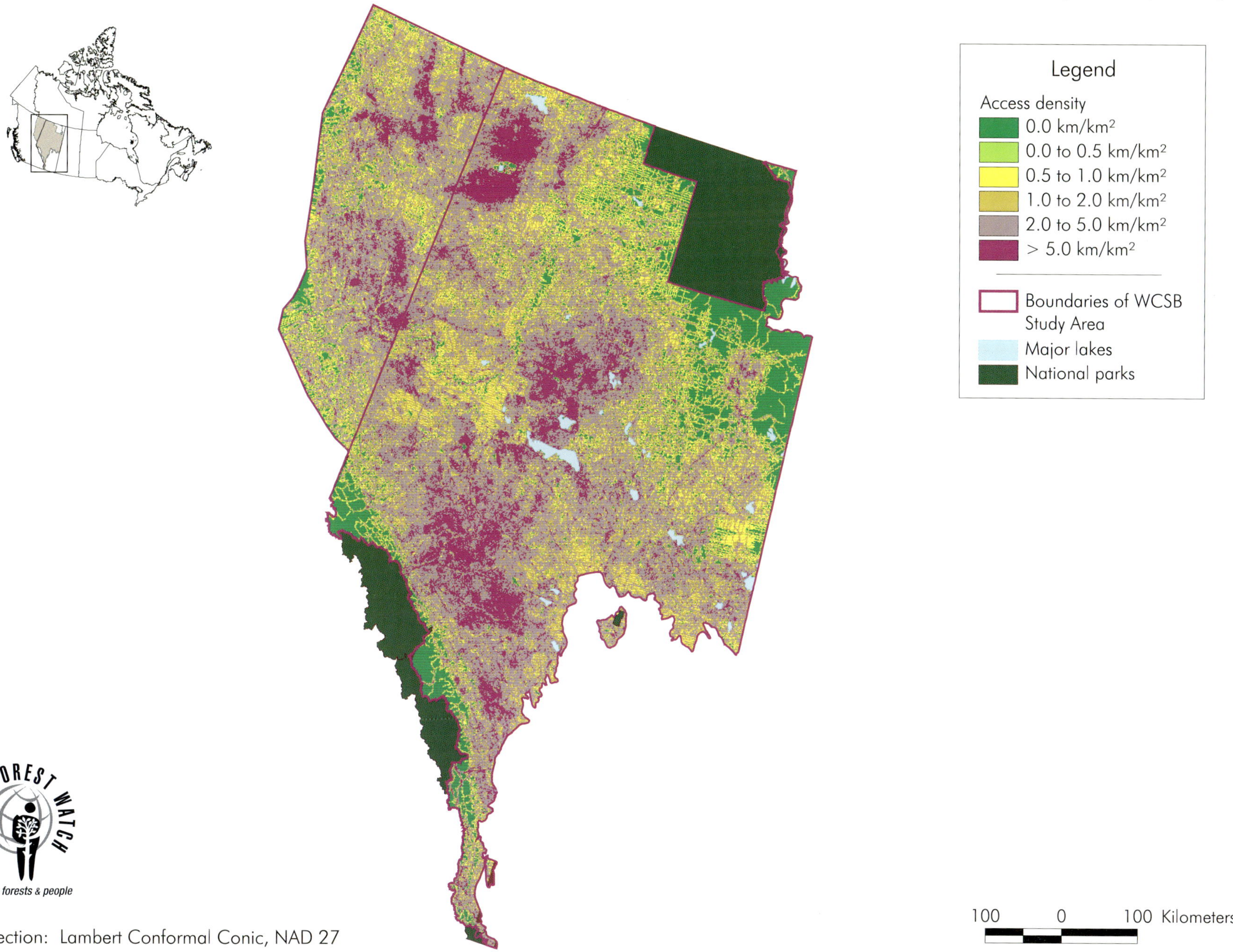

Linking forests & people

Projection: Lambert Conformal Conic, NAD 27

Scale: 1:7,000,000

Produced by the Headwaters Institute, Calgary for Global Forest Watch Canada

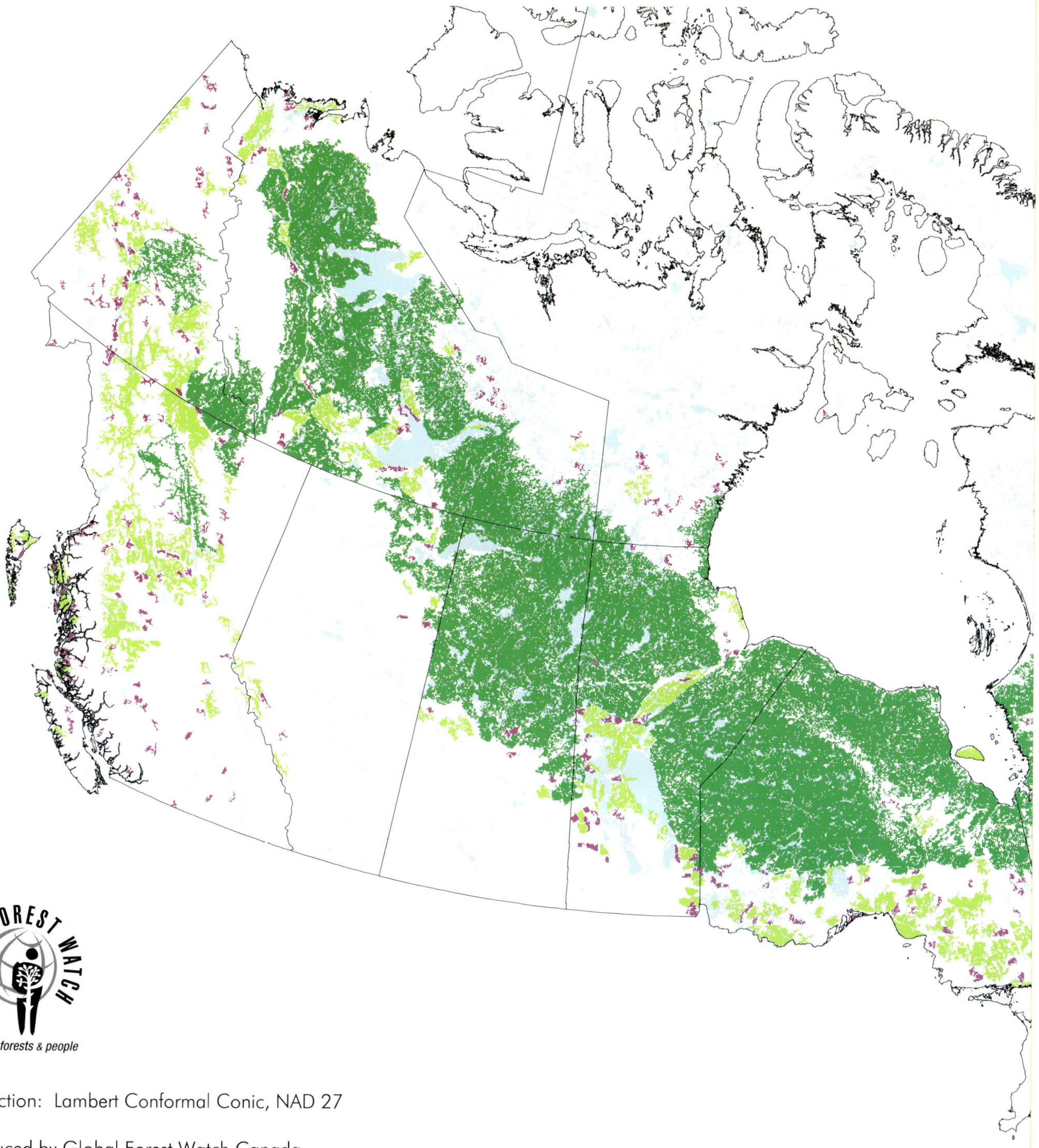

Projection: Lambert Conformal Conic, NAD 27

Produced by Global Forest Watch Canada

Birds are affected by loss or modification of habitat caused by logging, road development, climate change, and the spread of non-native species. From 1980 onward, more forest bird populations in Canada are showing declining populations compared with the period 1966-1979, perhaps as a result of changing forest conditions and other factors.[35]

Table 7. Canadian Forest-Dwelling Species at Risk, 1999

CATEGORIES	MAMMALS	BIRDS	PLANTS	REPTILES	TOTAL
Endangered	3	4	14	1	22
Threatened	2	4	11	3	20
Vulnerable	13	9	14	6	42
Total	**18**	**17**	**39**	**10**	**84**

Source: Canadian Forest Service, *The State of Canada's Forests: 1998-1999 Innovation* (Ottawa: Natural Resources Canada, 1999), p. 88.

SECTION 3. THE FOREST INDUSTRY

INTRODUCTION

Logging is the most widespread development activity within Canada's forests. With an annual log harvest of over 180 million cubic meters, the forest industry generates significant economic activity, export revenues, and jobs. In this section we look at the forest sector in more detail, examining the sector's economic value, the extent and rate of logging, and the environmental constraints to further cutting from a commercial forestry perspective.

The data presented here respond to three of GFW's monitoring goals. GFW 1) tracks existing and planned development activities, 2) provides data on actors—including the companies, individuals, government agencies, and others—engaged in this development, and 3) provides data on forest ecosystems to highlight the environmental and economic tradeoffs that development options entail.

In this section, we answer five specific questions related to these monitoring goals:

1. How important is the forest sector to Canada and the world?
2. How much of Canada's forest is allocated for logging?
3. What are current logging trends and how extensive are these activities?
4. Are these practices sustainable in terms of maintaining long-term timber production?
5. Who are the key actors involved in this activity? How do they benefit (or lose) from logging?

Summary of the Forest Industry

Canada is the world's leading exporter of wood products. The forest industry is a significant sector of the national economy, generating over $68 billion in sales in 1996 and accounting for about 3 percent of GDP. Three fourths of the timber harvest occurs in Quebec, Ontario, and British Columbia.

Over 220 million hectares of land are within volume- or area-based tenures (license agreements). Of this total, just under 167 million hectares are forested. Due to physical and other limiting factors, only a portion of this area is actually open to harvest. Tenures cover large areas of Canada—encompassing more than half of forests within 6 of the nation's 10 forest types.

To date, approximately one fifth of Canada's 235-million hectares of commercial forests have been logged. However, GFW analysis indicates that over half of the total commercial forest area faces productivity limitations because of climate, topography, and other factors. Currently, about 1 million hectares are cut annually. This adds to the cumulative impact of natural factors such as fire and insects, which play a major role in shaping Canada's forests.

Evidence suggests logging rates exceed long-term sustainable levels under current practices. The federal government estimates that 90 percent of all harvest occurs within primary and old-growth forests. This maximizes short-term yields at the expense of long-term production rates.

Key actors in the forest sector include forestry companies, communities, First Nations and Métis communities, and the public and consumers. As in other nations, the trend in the forest products industry is towards consolidation; 13 companies hold about 48 percent of commercial forest operations area. The forest sector generates about 350,000 direct jobs and over $11 billion in wages, particularly important in the 337 communities where the sector accounts for more than half the community's employment.

About 80 percent of Canada's 1 million First Nations and Métis live on reserves and communities in boreal or temperate forests. Although they hold claims to large areas of Canada's forest, these groups currently manage only a small portion of tenured areas. Management for timber production often conflicts with First Nations' rights and traditional holistic values toward forests.

Question 1 focuses on the overall economic significance of the forest industry to Canada. Questions 2, 3, and 4 address the magnitude and rate of logging and whether current practices can be maintained. Question 5 identifies the major actors, or stakeholders, in this industry.

Using data derived from the National Forestry Database Program (online at: http://nfdp.ccfm.org), Statistics Canada (online at: http://www.statcan.ca), and findings from several studies, we present 11 indicators to track key trends in the forest sector. We use this term loosely, since some indicators described here are actually a suite of measures that address a particular issue. These indicators are briefly described below.

Forest Industry Indicators

ECONOMIC VALUE OF THE FOREST INDUSTRY

Indicator 1 – Economic Value: We present data on production and export revenues, along with other measures that quantify the economic value of the forest industry.

Indicator 2 – Global Export Ranking: This is a measure of the significance of Canada's timber industry relative to other countries in terms of wood product exports. We also identify the major importers of these products.

LOGGING TRENDS

Indicator 3 – Forest Allocation: We present statistics on the extent of tenure allocation in Canada's forests, which indicates where logging development is underway. We also include figures on forest ownership.

Indicator 4 – Rate of Logging: This includes data on the extent and rate of logging in Canada's forests. With other measures, this is useful for addressing the sustainability of current practices in the logging industry.

NATURAL DISTURBANCE TRENDS

Indicator 5 – Insect Trends: Data on insects provide information on the overall health and condition of forests.

Indicator 6 – Fire Trends: Data on forest fire trends measure natural disturbance trends and the overall health of forests.

SUSTAINABILITY

Indicator 7 – Sustaining Long-Term Production: We include several measures useful for assessing whether current production levels can be maintained over the long term, including proportion of primary to secondary forest harvested, and production rates relative to AACs (government-set harvesting caps).

Indicator 8 – Regeneration: Data on restocking rates are helpful in assessing whether logged-over areas are regenerating and provide an indication of the intensity of management for timber production.

KEY ACTORS

Indicator 9 – Forestry Companies: We identify the major companies active in the forest industry and where they are operating.

Indicator 10 – Jobs and Wages: This indicator presents information on communities and people dependent on the timber industry.

Indicator 11 – First Nations and Métis: We include statistics on First Nations and Métis communities with a stake in the forests and logging development.

These indicators have limitations. They paint a picture of the value, nature, and sustainability of the forest industry, using measures quantifiable at regional and national scales with currently accessible data. They provide a baseline for further monitoring and reporting by GFW Canada. In the future, we hope to add additional measures that can more comprehensively depict trends in the logging sector. Data underlying the indicators also are limited by variations in age, data quality, and data collection methodologies. See Technical Notes and Appendix 1 for further details.

ECONOMIC VALUE OF THE FOREST INDUSTRY

Indicator 1: Economic Value

In 1996, there were more than 13,000 forest enterprises in Canada.[36] The forest industry consists of companies involved in forest management, logging, lumber production, pulp and paper, panel and board manufacturing, and engineered wood products. It also includes a wide variety of specialized secondary manufacturing or value-added operations.

The forest industry is a significant contributor to the Canadian economy

In 1996, the forest industry:

- Generated over $68 billion (US $47 billion) in sales.[37]
- Contributed $8.9 billion (US $6.1 billion) in taxes and other payments.[38]
- Paid over $11 billion (US $7.6 billion) in wages.[39]

The forest sector contributes 3 percent to GDP

In 1998, the forest sector contributed approximately $18.2 billion (US $12.4 billion) to the nation's gross domestic product (GDP),[40] or about 2.5 percent of the total.[41] The sector's contribution to GDP remained about 3 percent per year during the 1990-1996 period.[42]

Forest-product exports generated almost $40 billion (US $27 billion) in sales in 1998

Forest-product exports account for over 10 percent of Canada's total exports

Forest-product exports are a significant portion of Canada's economy in relation to other sectors. *(See Figure 4.)* Forest products accounted for almost $32 billion (US $22 billion) in trade surplus (exports minus imports) in 1998.[43] The sector's contribution to the overall balance of trade has remained above 10 percent for the last 5 years.

Figure 4. Net Trade for Various Sectors in Canada, 1997

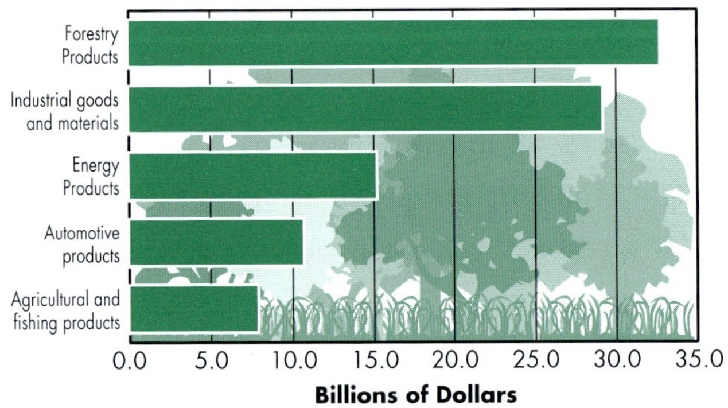

Sources: Statistics Canada, "Exports of goods on a balance of payments basis." Online at: http://www.statcan.ca/english/PGDB/Economy/International/gblec04.htm (January 17, 2000). Statistics Canada, "Imports of goods on a balance of payment basis." Online at: http://www.statcan.ca/english/PGDB/Economy/International/gblec05.htm (January 17, 2000).

Indicator 2: Global Export Ranking

Canada is the world's leading exporter of timber products

Canada was the biggest forest-products exporter by value in the world in 1996.[44] Canada exports nearly 20 percent of the total global value of all forest products. Overall, Canada's forest-product exports have increased since 1991.

Four fifths of all forest products are exported to the U.S.

Canadian forest products are exported to a number of countries. *(See Figure 5.)* The U.S. (79 percent), Europe (8 percent), and Japan (7 percent) are the most important markets for Canadian forest products.[45] These products include softwood lumber, paper and paperboard, wood pulp, and newsprint.[46] The major exporting provinces are British Columbia with $13.2 billion (US $9.1 billion), Quebec with $10.8 billion (US $7.5 billion), and Ontario at $8.1 billion (US $5.6 billion).[47]

Figure 5. Importers of Canadian Forest Products

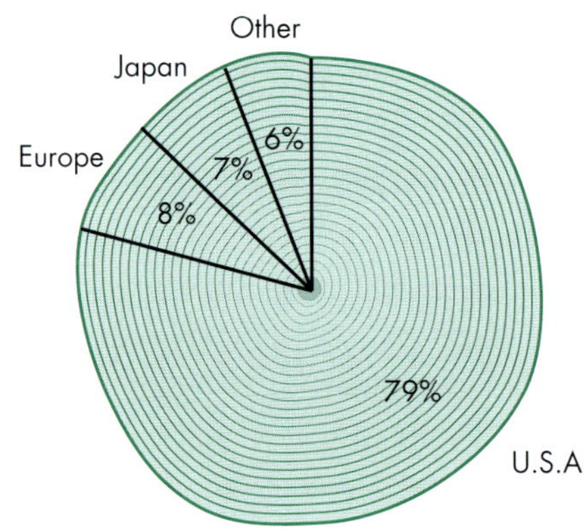

Source: Canadian Forest Service, *The State of Canada's Forests: 1998-1999 Innovation* (Ottawa: Natural Resources Canada, 1999), p. 26.

LOGGING TRENDS

Indicator 3: Forest Allocation

94 percent of forests are publicly owned

Some 71 percent of forests are controlled and managed by the provincial governments. The federal government owns 23 percent.[48] This land is held and managed by governments on behalf of the public. The remaining 6 percent of forests, or 25.1 million hectares, are held by private landowners.[49] The provinces and the Northwest Territories have their own legislation, regulations, and programs governing the allocation of public forest logging rights and management responsibilities.

Most provinces have entered into long-term tenures or license agreements, which give rights to companies to log forests. Tenures are of two principal types:

- Area-based tenures are usually 20 to 25 years in duration. The holder has responsibility for harvesting and forest management for a specific area. In exchange, companies pay stumpage, rents, and/or royalties. Increasingly, companies are also undertaking other management responsibilities such as replanting logged areas.[50] The operational responsibilities—reforestation, protection, and road building—for government and industry differ based on the jurisdiction and type of tenure.
- Volume allotments are typically of shorter duration and may not require the holder to be responsible for forest management.

Each province controls the logging rate on Crown (public) land through the determination of an allowable annual cut (AAC). The AAC is the maximum amount of timber that can be cut annually from a specified area over a given period of time.[51] The determination of AACs varies considerably from province to province.[52] AACs are generally recalculated every 5 to 10 years. A range of factors are considered, including the extent of the forest area, growth rate of trees, losses due to fire, accessibility, jobs, and economic and government revenue targets.[53] Critics say the process of setting AACs lacks a national standard and transparency.[54] Recognition of other forest values is also beginning to influence AAC determination.

More than 50 percent of Canada's forests are within the commercial forest zone

42 percent of Canada's forests are in volume- and area-based tenures

More than 286 million hectares are within the commercial forest zone, which includes tenure

Technical Note 7. Tenure

The commercial forest zone includes both area-based tenures and area-approximated tenures (from volume-based AACs for British Columbia and Quebec), plus Forest Management Units (where quotas are given) and proposed new tenures. These numbers should be considered approximate; they are a compilation of various types of allocation systems. We provide results based on harvesting allotments that coincide with forest cover areas as well as total areas. Our data is based on World Wildlife Fund Canada tenure data that we have expanded and updated. *(See Appendix 1.)*

Table 8. Tenure by Forest Region

FOREST REGION	COMMERCIAL FOREST ZONE (000 HECTARES)	FOREST IN THE COMMERCIAL FOREST ZONE (000 HECTARES)	VOLUME- AND AREA-BASED TENURES (000 HECTARES)	FOREST IN TENURES (000 HECTARES)	FOREST IN TENURES (PERCENT)
Boreal	164,523	126,457	125,430	101,845	47
Aspen Parkland	7,836	330	607	64	2
Taiga	17,203	10,923	2,741	1,782	2
Subalpine	20,184	12,041	18,950	11,326	68
Montane	13,003	10,716	12,723	10,658	88
Coast	8,914	5,856	8,914	5,856	79
Columbian	4,842	3,419	4,842	3,419	88
Great Lakes	20,658	18,990	19,505	18,015	65
Acadian	10,345	8,803	10,345	8,803	96
Forests outside Major Forest Regions	18,677	5,286	16,243	5,067	28
All Forest in Canada	**286,185**	**202,821**	**220,300**	**166,835**	**42**

Sources: GFW Canada.
Note: We have included all forest cover in Canada for this analysis, including that found in tundra and grassland. Forests in the Carolinian Forest Region are not included as there are no tenures in this region.

areas and forest management units. *(See Technical Note 7.)* Of that total, more than 202 million hectares are forest. More than 50 percent of Canada's forests are within the commercial forest zone. (This percent is based on the 402 million hectares of forest in classes 1-12 in the *Land Cover of Canada* map, rather than the 417.6 million hectares from the Canadian Forest Service data). *(See Table 8.)*

Approximately 220 million hectares of the 286 million hectares are in volume- and area-based tenures. Of this total, approximately 167 million hectares (42 percent) are forested.

Note that the 50 percent figure presented here differs slightly from the amount presented in Section 2 (52 percent) because of the methodology used in that section to calculate cumulative extent of development.

6 of 10 Forest Regions have more than half of their forests in tenures

Within four regions, tenures cover more than 75 percent of the forest

Tenures and forest management units predominate in most forest regions. Allocation levels to tenures alone are very high in many regions, including the Acadian Forest Region (96 percent), Columbian Forest Region (88 percent), and Montane Forest Region (88 percent). Over two thirds of the Subalpine Forest Region (68 percent) and Coast Forest Region (79 percent) are in some type of timber allotment system. Tenures cover almost 50 percent of the Boreal Forest Region.

British Columbia, New Brunswick, and Nova Scotia each have over 80 percent of their forests allocated

British Columbia, New Brunswick, and Nova Scotia have the highest proportion of their forests tenured. Ontario has 60 percent of its forests allocated, Quebec 43 percent, Alberta 45 percent, Manitoba 39 percent, and Saskatchewan 33 percent. *(See Table 9.)*

Map 6 illustrates cutblocks approved for logging in the various five-year Forest Development Plans (for the years 1998 to 2003) for a portion of British Columbia's temperate rainforest, sometimes referred to as the "Great Bear Rainforest." This data was compiled by Forest Watch of British Columbia and its partners from government and industry planning documents and took 2 years to complete. Data of this type and on this scale is not readily available regionally or provincially in Canada, although it is essential data for accurately tracking where forests have been logged. (Subsequent amendments to these various Forest Development Plans are not represented.)

Table 9. Tenure by Province

PROVINCE	COMMERCIAL FOREST ZONE (000 HECTARES)	FOREST IN THE COMMERCIAL FOREST ZONE (000 HECTARES)	VOLUME- AND AREA-BASED TENURES (000 HECTARES)	FOREST IN TENURES (000 HECTARES)	FOREST IN TENURES (PERCENT)
Alberta	42,985	25,474	21,000	16,842	45
British Columbia	80,382	49,212	80,382	49,210	82
Manitoba	34,325	20,029	14,131	9,534	39
New Brunswick	5,964	4,926	5,964	4,926	86
Newfoundland	12,067	7,268	3,998	2,349	10
Nova Scotia	5,167	4,315	5,167	4,315	~100
Ontario	52,620	44,902	38,580	34,321	60
Quebec	41,260	36,315	41,260	36,310	43
Saskatchewan	13,348	10,790	11,549	9,383	33
Yukon Territory	71	69	70	69	0.3
Total	**288,189**	**203,300**	**222,101**	**167,259**	**47**

Sources: GFW Canada.
Note: Provincial areas are for land area that overlaps with the Land Cover of Canada map only.
Figures do not equal totals in Table 8 because of the effects of coarse provincial boundaries.

Indicator 4: Rate of Logging

The rate of logging has increased substantially throughout the 20th Century. *(See Figure 6.)* The current annual rate of logging—close to 1 million hectares—equates to a harvest of 0.4 percent of Canada's commercial forest base of 235 million hectares,[55] or 0.8 percent of the 119 million hectares currently managed primarily for timber production.

Harvest rates grew 60 percent between 1975 and 1988

Between 1975 and 1988, the amount harvested rose 60 percent to a peak of approximately 1.1 million hectares in 1988.[56] Logging levels fluctuated in the 1990s, rising again to 1.02 million hectares in 1997, the most recent year for which national data are available.[57]

About one fifth of commercial forests have been logged to date

Maritime provinces have the highest proportion of logged forest

From 1975 to 1996, logging occurred on approximately 8.4 percent of Canada's commercial forest lands.[58] *(See Figure 7.)* Although cumulative harvest data are not available before 1920, GFW Canada estimates that no more than 50 million hectares of Canada's forests have been logged to date, representing approximately 23 percent of Canada's commercial forests. This estimate does not mean that 23 percent of the country's primary forest has been logged. Some forests have been logged twice or more, particularly in eastern Canada and the Maritimes, where there is a long history of logging activity. Statistics are not readily available on the amount of primary versus secondary forest logged. Environment Canada reports that 90 percent of logging takes place in areas not previously cut commercially.[59] If this figure applies historically, it implies that roughly 45 million hectares of primary forest (about 20 percent of Canada's commercial forests) have been logged to date.

Most forests are clearcut

Clearcutting has been, and continues to be, the dominant harvesting practice. Between 1920 and 1996, approximately 42 million hectares (88 percent) were clearcut.[60] The remaining 12 percent was logged using selection methods. The appropriateness of clearcutting is a matter of significant public debate in Canada. *(See Box 5.)*

Figure 6. Area Logged Annually in Canada, 1920-1995

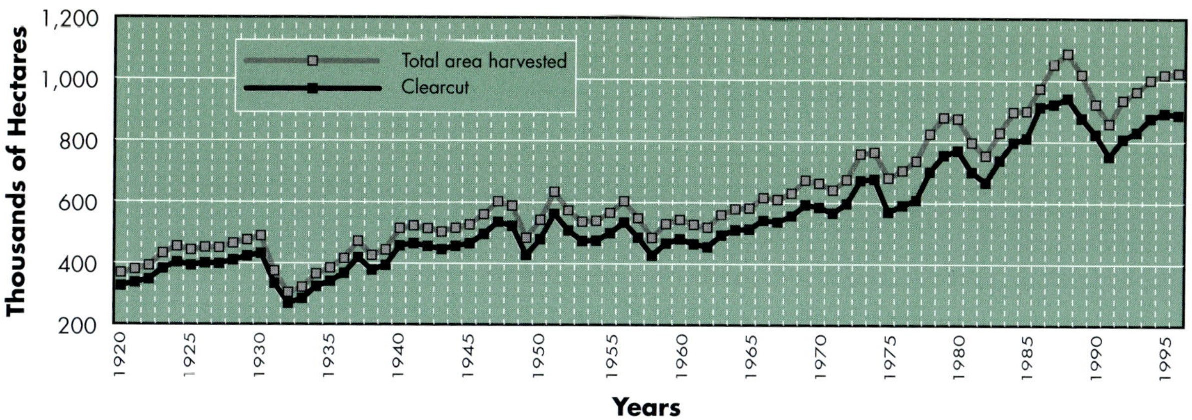

Sources: (For data from 1920-1992) Environment Canada, *Sustaining Canada's Forests: Timber Harvesting, Technical Supplement No. 95-4* (Ottawa: Environment Canada, 1995). Online at: http://www3.ec.gc.ca/~ind/English/Home/default1.htm (January 19, 2000).
(For data from 1993-1996) Canadian Council of Forest Ministers, "National Forestry Database Program: Silviculture: Table 6.1.1." Online at: http://nfdp.ccfm.org (January 19, 2000).

Figure 7. Productive Forest Land Harvested, 1975-1996

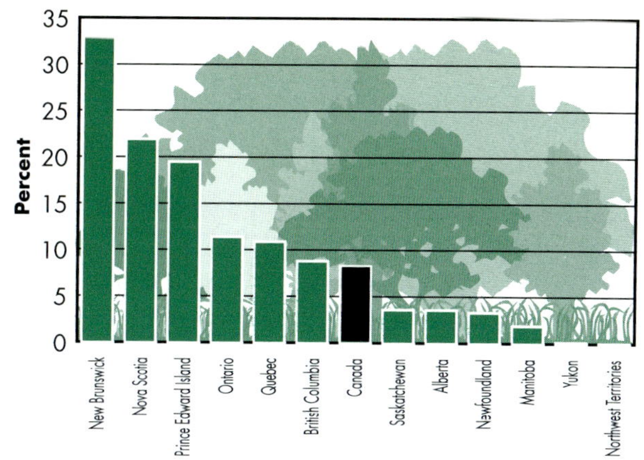

Source: Canadian Council of Forest Ministers. "National Forestry Database Program: Silviculture and Forest Inventory." Online at: http://nfdp.ccfm.org (January 19, 2000.)

Almost three quarters of the harvest occurs in British Columbia, Quebec, and Ontario

By volume, annual harvest rates are highest in British Columbia

Quebec logs more hectares than any other province

The provinces of Quebec, Ontario, and British Columbia accounted for almost three fourths (72 percent in terms of area and 73 percent in terms of volume) of all logging in 1997. *(See Table 10.)*

The amount of wood harvested per hectare varies, ranging from an average of 95 m^3/ha to 438 m^3/ha. These numbers are averages per province and mask much higher and lower numbers within provinces. For example, volumes of more than 700 m^3/hectare are possible in the coastal rainforests in British Columbia. The variability between provinces is related to differences in productivity of forest types and to other factors such as age distributions of forest stands.

Table 10. Area and Volume Logged by Province, 1997

PROVINCE	AREA (000 HECTARES)	TOTAL AREA (PERCENT)	VOLUME (m^3)	TOTAL VOLUME (PERCENT)	VOLUME/ HECTARE (m^3/ha)
Quebec	364	36	41	22	111
Ontario	198	19	25	13	124
British Columbia	176	17	69	38	394
New Brunswick	112	11	11	6	100
Nova Scotia	69	7	7	4	95
Alberta	51	5	22	12	438
Newfoundland	20	2	2	1	100
Saskatchewan	18	2	4	2	234
Manitoba	16	2	2	1	135
Northwest Territories	4	0	NA	NA	NA

Source: Canadian Forest Service, *The State of Canada's Forests: 1998-99 Innovation* (Ottawa: Natural Resources Canada, 1999), pp. 26-32. Online at: http://nrcan.gc.ca/cfs/proj/ppiab/sof/common/latest.shtml (January 18, 2000).
Note: Area and volume numbers are rounded.

BOX 5. The Debate on Clearcutting

The appropriateness of clearcutting versus other silvicultural practices is a matter of considerable public debate and concern. Few generalizations, however, can be made that would apply to all of Canada's forests. In ecological terms, silvicultural practices can be assessed in terms of how well they mimic natural patterns of disturbance. Natural disturbance processes vary greatly across Canada. For example, fire plays an important role in shaping the structure and composition of boreal forests. Fire cycles have durations of 40 to 60 years for low-intensity ground fires to over 200 years for stand-replacing fires.[1] In contrast, large-scale natural disturbances are rare in the coastal temperate rain forest. In Clayoquot Sound, on the west coast of Vancouver Island, large-scale natural disturbances such as extensive blowdowns or wildfires are often of low intensity and recur only after long intervals of 400 to 1,000 years or more.[2] As a result, landscapes are dominated by old-growth forests, with individual trees reaching an age of almost 1,000 years.[3]

Fire Disturbance and Forest Harvest Policy in Ontario

The appropriateness of clearcutting as a logging practice has been the subject of lengthy public discussions in Ontario and remains unresolved. After years of debate and public hearings, the Ontario Environmental Assessment Board mandated in 1994 that, save for exceptional circumstances, clearcuts could not exceed 260 hectares in size, and that the size restriction should not be avoided though odd configurations or contiguous cuts.[4] This decision made large contiguous clearcuts illegal in Ontario, subject to limited exceptions.

The ruling appears to be contradicted by the Ontario Ministry of Natural Resources' (MNR) interpretation of Ontario's new forest legislation, the *Crown Forest Sustainability Act* (CFSA), which was passed in 1994. The Act states that:

> The long term health and vigour of Crown forest should be provided for by using forest practices that, within the limits of silvicultural requirements, emulate natural disturbances and landscape patterns while minimizing adverse effects on plant life, animal life, water, soil, air, and social and economic values, including recreational values and heritage values.[5]

Five years later, MNR appears to have interpreted its new legislation to endorse an abandonment of the Environmental Assessment Board size limit on clearcuts. As a result, an internal report calls for clearcuts of up to 10,000 hectares.[6] MNR field staff claim that such clearcuts are necessary to meet the requirements of the CFSA because they will mimic natural disturbance, most notably that of fire. Recent developments in the forests of Temagami illustrate the course of this new direction. The recently approved 1999-2019 Temagami Forest Management Plan contains 19 clearcuts larger than 260 hectares.[7] Several other forest plans under development also contain large clearcuts that exceed the Environmental Assessment Board limit.

The government of Ontario appears to be pursuing an approach to forest management based on the assumption that clearcuts mimic fire. They contend that because small, medium, and large fires are a natural part of the boreal forests of Ontario, clearcuts should also be of various sizes, including very large openings up to 10,000 hectares. However, the available science indicates that clearcuts do not mimic fire disturbance. More scientific inquiry and public debate will likely be needed.[8]

Changing Forest Management Practices on the British Columbia Coast

In June 1998, MacMillan Bloedel (acquired by Weyerhaeuser in November 1999) announced it would gradually phase out clearcut logging on its 1 million hectares of private and Crown forest land in Coastal British Columbia. MacMillan Bloedel introduced a variable retention harvesting regime, which would retain biological legacies such as snags and veteran trees in the stand after logging, as well as a system of stewardship zones at the landscape level.

MacMillan Bloedel's efforts have drawn qualified support from scientists and environmental nongovernmental representatives, particularly for changes made at the stand level. But concerns remain with respect to adequate protection of old-growth forest at the landscape level, riparian protection, and the overall level of cut on MacMillan Bloedel's lands, which is not projected to decline substantially.[9]

[1] R.T. Graham and T.B. Jaim. 1998. "Silviculture's role in managing boreal forests." *Conservation Ecology* 2(2): 8-29.

[2] Clayoquot Sound Scientific Panel, *Sustainable Ecosystem Management in Clayoquot Sound: Planning and Practices* (Victoria: Crown Publications, 1995), p. 21.

[3] Clayoquot Sound Scientific Panel, *Sustainable Ecosystem Management in Clayoquot Sound: Planning and Practices* (Victoria: Crown Publications, 1995), p. 23.

[4] Ontario Environmental Assessment Board, *Class Environmental Assessment by the Ministry of Natural Resources for Timber Management on Crown Lands in Ontario No. EA 87-02* (Toronto: Ontario Environmental Assessment Board, 1994).

[5] 35th Legislature, Ontario, *Bill 171—An Act to revise the Crown Timber Act to provide for the sustainability of Crown Forests in Ontario* (Toronto: Legislative Assembly of Ontario, 1994).

[6] Ontario Ministry of Natural Resources, *Forest Management Guidelines for the Emulation of Fire Disturbance Patterns -Analysis Results* (Toronto: Ontario Ministry of Natural Resources, 1997).

[7] Ontario Ministry of Natural Resources, *Forest Management Plan for Temagami Management Unit 1999-2019: Plan Summary* (Toronto: Ontario Ministry of Natural Resources, 1999).

[8] T.J Carleton, P. MacLellan. 1994. "Woody vegetation response to fire versus clear-cutting: a comparative survey in the Canadian central boreal forest." *Ecoscience* 1(2): 141-152.

[9] Dovetail Consulting Inc., *Summary of First Year Critique workshop on the MacMillan Bloedel BC Coastal Forest Project, July 14-16, 1999* (Vancouver: MacMillan Bloedel Limited, 1999).

NATURAL DISTURBANCE TRENDS

Forest fires and insect infestations are a natural part of the forest life cycle, helping to renew forest ecosystems and shape forest biodiversity. Forests are well adapted to disturbances that correspond to natural "background" levels of frequency and intensity.

Trends in the intensity and frequency of fires, insect outbreaks, and other natural disturbances are relevant to timber production in several ways. First, they show the proportion of timber stocks precluded from harvest due to natural damage. Second, in theory they also can be used to indicate the effectiveness of management interventions designed to prevent such losses. However, it can be difficult to determine whether trends are due to management interventions or natural cycles.

Third, disturbance trends are an indication of forest health. Compared to disturbance patterns in forests where human activities are minimal, they indicate how closely management regimes approximate natural conditions. For example, suppression of natural disturbances such as fire can have negative consequences on species that recolonize burnt areas and encourage build-up of flammable debris, which may ultimately result in catastrophic fires. Ideally, a forest managed both for timber and nontimber values would exhibit natural disturbance patterns closely approximating those of a forest undisturbed by commercial-scale human activities.

Indicator 5: Insect Trends

Insect outbreaks are in decline

Like fire, insects and disease are a natural part of Canada's forests. Two common forest insects that attack commercial forests in Canada are the spruce budworm *(Choristoneura fumiferana)* and the tent caterpillar *(Malacosoma disstria)*. Spruce budworm is common to boreal forests, but infestations have been most severe in Atlantic Canada, where extensive forest harvesting created conditions suitable for a budworm outbreak. The forest tent caterpillar is widely distributed across Canada and is a serious pest of trembling aspen. In western Canada, the mountain pine beetle *(Dendoctonus ponderosae)* attacks mostly lodgepole pine in the montane forests of Alberta and British Columbia.

Annual populations of these species vary widely in response to environmental factors. During the 1976-95 period, the total forest area affected by moderate to severe defoliation by insects in Canada declined sharply from approximately 65 million hectares to 7 million hectares, largely as a result of declines in spruce budworm populations. *(See Figure 8.)* The area of forest moderately to severely affected by spruce budworm populations has declined 97 percent over this period. Over 5 million hectares were moderately to severely affected by insect defoliators in 1998.[61]

Figure 8. Forest Affected by Insects (Defoliation and Dead Trees), 1976-1996

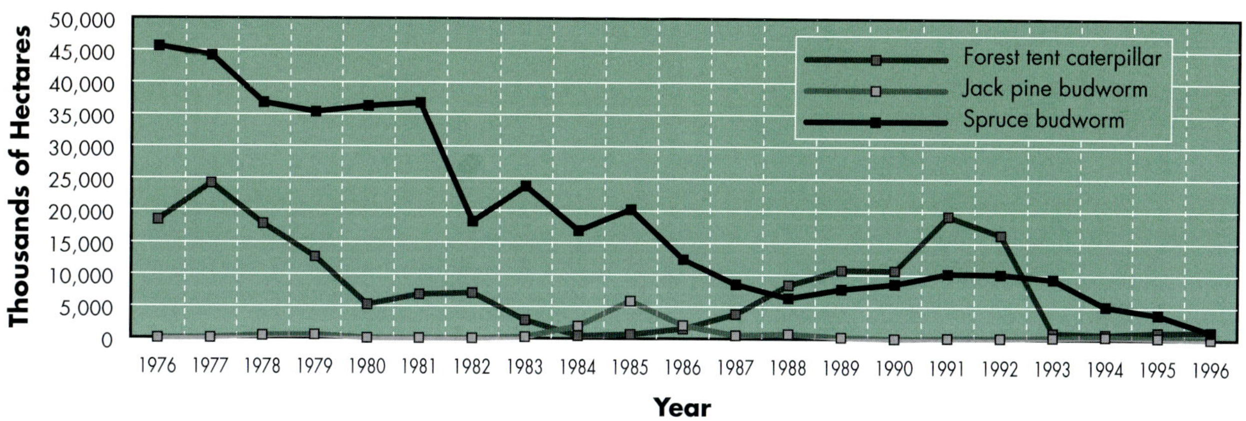

Source: Canadian Council of Forest Ministers. "National Forestry Database Program: Forest Insects and Diseases." Online at: http://nfdp.ccfm.org (February 7, 2000).

Indicator 6: Fire Trends

The overall trend in Canada has been towards more frequent and larger fires in recent decades. For example, the number of fires recorded between 1960 and 1995 was 60 percent higher than the total for 1920 to 1960.[62] It is important to note that data for earlier years may be less complete than more recent data.

In an average year during the 1970–97 period, 9,000 fires burned more than 2.1 million hectares of forest.[63] Most of these fires were small. In fact, 91.5 percent were less than 10 hectares in size; only 1.5 percent exceeded 1,000 hectares. But the few larger fires accounted for 93.1 percent of the total area burned in Canada.

Approximately 80 percent of the area burned is considered commercially nonproductive.[64] On average, the burned area of productive stocked timber is lower than the area harvested for wood. *(See Figure 9.)*

Figure 9. Area Burned by Forest Fires, 1970-1995

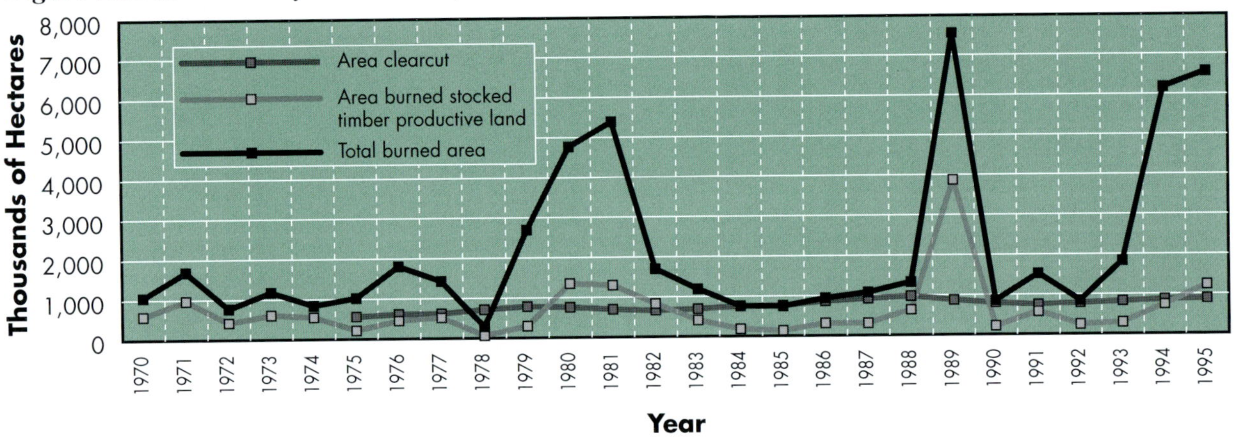

Source: Canadian Council of Forest Ministers. "National Forestry Database Program: Forest Fires." Online at: http://nfdp.ccfm.org (February 7, 2000).

SUSTAINABILITY

Indicator 7: Sustaining Long-Term Production

To address the issue of whether current harvest rates can be maintained over the long term, we look at several measures. These include harvest rates relative to AAC (government-set harvesting levels); productivity limitations within commercial forests; regeneration trends; and insect and fire outbreaks, which can reduce the amount of timber available for harvest.

Taken together, these measures suggest that while more intensive management has resulted in improved regeneration rates, current harvest rates are maintained by cutting extensive areas of primary and old-growth forest. There are constraints to expanding operations into new areas, as most tenured areas face some form of productivity limitations for commercial forestry. Furthermore, tenures already extend into far northern forests, which require very long periods of time to regenerate.

Harvest rates are generally below AAC cut levels

AAC cut levels have traditionally been based on maximizing timber production

Actual logging rates in Canada have been below the national AAC each year for the period from 1970 to 1996, suggesting that timber supply is sustainable.[65] However, many Canadian AACs have been deliberately set above long-term sustainable harvest levels in order to log extensive primary and old-growth forests, which yield higher timber volumes.[66] For example:

- In British Columbia, current AACs exceed long-term sustainable harvest levels by 11 million cubic meters per year, or 18.5 percent.[67]
- 90 percent of Timber Supply Areas (government-administered forest management units) in British Columbia are being logged above long-term sustainable levels; one in three by more than 50 percent.[68]
- In Alberta, remaining old-growth forests on Crown (public) land will be liquidated in about 12 years for forests older than 150 years, and about 41 years for forests older than 120 years.[69]

Traditionally, Canada's forests have been described by foresters as skewed to older age classes. These older age classes are considered overmature and a potential waste of timber resources, as they will deteriorate and die if not logged. They "must be cut... to attempt to approach a 'normal' forest."[70] This "normal" forest refers to a more balanced age distribution that offers better long-term yields from a harvesting perspective and not to any ecological or biologically "normal" state.

Current rates of timber harvest are expected to fall significantly in order to meet several sustainable forest management objectives such as conservation of biological diversity and recreation opportunities.[71] Case studies estimate that the rate of logging in Canada would have to decline by 10 to 25 percent in the boreal forests of Canada[72] and 30 to 40 percent on the coast of British Columbia[73] to address broader forest sustainability objectives.

In some areas, harvesting rates have reached AAC levels. In other areas, timber supply shortages have been reported. While Canada does not face a shortage of wood supply, it does face shortages of softwood timber in the near future.[74] Harvest of softwood species (e.g. pine and spruce), which accounted for more than 86 percent of Canada's commercial logging in 1995, approached the total AAC for those species that year.[75]

Almost a third of Canada's forests were logged or affected by fires and insects between 1990 and 1996

When natural disturbance is factored in, the percentage of area logged relative to area available for harvest rises significantly

To ensure forest regeneration, appropriate harvest rates need to factor in the total area of forest disturbed by logging as well as natural factors. Official statistics indicate that less than 1 percent of commercial forest area is harvested annually. But this does not include fire and insect damage, which affects a considerable portion of the commercial forest.

Cumulatively between 1980 and 1996, 69 million hectares of forest were affected by insects, 43 million hectares by fire, and 16 million hectares by harvesting. It is difficult to calculate the total amount of forest affected, since areas may have been subject to more than one of these factors. The 69 million hectares affected by insects is equal to 16.5 percent of the entire forest area in Canada. Assuming no overlap, 128 million hectares (30 percent of all forests) were affected by logging or natural disturbance over this 16-year period. The reality on the ground is somewhere in between these two figures.

Within the 235 million hectares of commercially productive nonreserved forest, 16 million hectares were logged and 12.7 million hectares subject to fires between 1980 and 1996. This corresponds to a total of 28.7 million hectares (or 12 percent) of forest theoretically available for timber harvest. Although there may be overlap once again between areas burned and logged, it does not factor in insect damage. Because data are not available on fire and insect damage occurring within the 119 million hectares of forest currently managed for timber production, it is difficult to estimate what proportion of these lands are disturbed annually both from logging and from natural events. Such information would provide a more accurate measure of whether current management regimes are sustainable.

Canada is clearing primary and old-growth forest to maintain production levels

Logging historically has occurred in the more productive southern forests of Canada, where extensive areas are now second-growth forest. In some parts of Canada, second- or third-rotation logging has occurred. However, as noted above, government figures indicate 90 percent of logging occurs within primary and old-growth forest—forests of high biodiversity and wilderness value. Each year Canada has a shrinking supply of "old-growth" or primary forests."[76]

Outside the southernmost forests, the effects of logging can be shown by comparing age-class distribution data for accessed and non-accessed forest stands. In the Canadian Shield portion of the Boreal Forest Region, which lies across northern Ontario, Quebec, and Newfoundland, trees older than 100 years occupy 18 percent of accessed areas, while they occupy 40 percent of the unaccessed areas.[77] (Note that forests within this area are rarely older than 140 years due to the relative frequency of fires). Given that this area is now largely allocated to timber production, it raises questions about harvest rates and AACs, especially if nontimber values such as biodiversity and carbon storage are considered. Within extensive boreal forest areas of Canada, these data suggest a shift from mature to younger forest as a result of logging.

The logging frontier is expanding into slow-growing, far-northern forests

With much of the southernmost forest logged, timber companies are expanding their operations northward into increasingly marginal timber areas.[78] In recent decades, provincial governments have issued forest tenures within extensive areas of Canada's northern, previously undeveloped forests. This trend appears to be continuing. In a June 1999 report on boreal forests issues, a subcommittee of the federal Standing Senate Committee on Agriculture and Forestry reported that western boreal forest allocations are a recent development. Excluding parks, Manitoba has allocated 62.8 percent, Saskatchewan 34.3 percent, and Alberta 70.2 percent of their boreal forest area to forest companies.[79] There is considerable public concern about whether this is sustainable, given the increasing limitations to commercial forestry at higher latitudes.

About 33 percent of tenured forest lands face severe productivity limitations

A further 28 percent face moderate limitations

To provide a rough estimate of the total area of Canada's forest tenure facing productivity limitations, Global Forest Watch Canada compared maps of the Commercial Forest Zone and Canada Land Inventory (CLI) forest capability classes. The latter map, while dated, provides the only existing national dataset depicting areas where there are constraints to timber harvesting *(See Technical Note 8 and Map 6.)*

Of the land within the Commercial Forest Zone:

- 28 percent have moderate limitations for the growth of commercial forests;
- 33 percent have severe to very severe limitations; and
- 19 percent are considered nonforested or unsuitable for commercial tree growth. *(See Figure 10.)*

These numbers suggest that extensive areas of forest currently reported by governments as within the commercial forests may be unsuitable for commercial forestry. As mentioned earlier, this does not mean that these areas will necessarily be logged. All forest tenure areas contain areas of low productivity that are routinely excluded from the operable land base. For example, only 25 percent of Alberta Pacific's 58,000 km^2 Forest Management Area is managed for timber production, while 59 percent is wetlands and 5 percent is set aside as reserves.[80]

It is also true, however, that the operable land base is subject to change. As prices rise or new technologies are introduced, formerly inoperable areas become economic to log. If a significant proportion of projected future timber supply needs of the industry must be obtained from old-growth or primary forests, rather than second growth, there may be pressure to log increasingly marginal forests to maintain current logging levels. Box 6 provides examples.

Figure 10. Tenured Forests with Limitations to Commercial Forestry

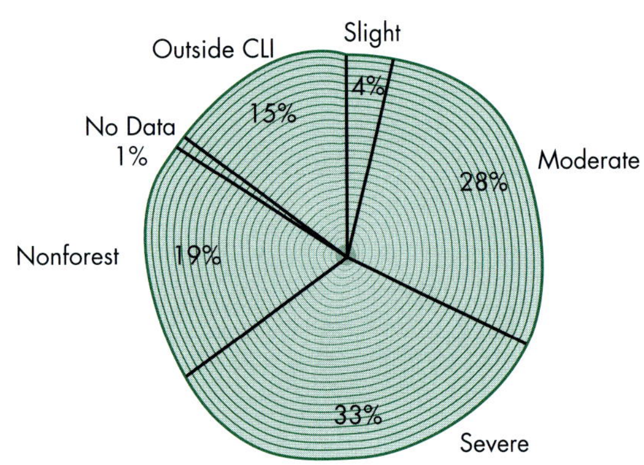

Source: GFW Canada

Technical Note 8. Ecological Limitations to Commercial Forestry

GFW Canada undertook a mapping analysis to assess the limitations to commercial forestry within Canada's Commercial Forest Zone. GFW Canada's analysis is based on the Canada Land Inventory (CLI), a detailed assessment of the ability of land to support commercial forestry based on soils, climate, landform, and vegetation.[1] The CLI was developed by the federal government to provide a basis for land use planning and covers approximately 25 percent of Canada's landbase. The CLI uses 7 land classes based on the physical limitations of the land to grow trees. The productivity associated with each class is based on the mean annual increment of the native tree species best adapted to the site at or near rotation age.

Since the dataset is 30 years old and the definition of "commercially viable" can change, we conducted a comparison with current British Columbia Ministry of Forest data to verify the accuracy of this data. We took a section of the CLI data that corresponds with the Cranbrook Timber Supply Area in interior B.C. The results show that there is not a great variation between classes in the two different datasets. There are major differences, however, in the terms used to describe each class. See our website for more information: www.globalforestwatch.org.

[1] Canada Department of Forestry and Rural Development, *The Canada Land Inventory Land Capability Classification for Forestry, Report No.4.* (Ottawa: Canada Department of Forestry and Rural Development, 1967).

BOX 6 Examples of Allowable Annual Cut (AAC) Issues

Williams Lake Timber Supply Area

The Williams Lake Timber Supply Area (TSA) is located in the Caribou-Chilcotin region in the central interior of British Columbia. This 4.9-million-hectare region has an AAC of 3.8 million m^3 per year, which is 63 percent above the long-term sustainable level of 2.4 million m^3 per year.[1] In recent years, in order to maintain logging levels, licensees have increasingly accessed "problem forest types"—stands of poor quality or low volume that traditionally have been uneconomical to log. In addition to demands being placed on the "problem forest types" to meet the conventional sawlog AAC, logging is also planned in these areas to support pulpwood agreements. Currently, approximately 200,000 m^3 (5 percent) of the AAC for the Williams Lake TSA is from "problem forest types." The region, similar to many regions in British Columbia, faces an acute timber supply shortfall over the next few decades.

James Bay Region, Quebec

In recent years, the Quebec government has authorized the expansion of logging into northwestern Quebec through the issuance of long-term tenure agreements, known as Timber Supply and Forest Management Agreements (TSFMA). The Region 10 area around James Bay accounts for 15–17 percent of Quebec's output by volume. The area includes approximately 130,000 km^2 of forested land (primarily open black spruce woodlands), of which 63 percent is considered productive forest land. The annual allowable cut is thought to be approximately 5 million m^3/year.[2] These forested lands are also home to approximately 6,000 Cree aboriginal peoples. The rapid extension of logging, involving clearcutting of 50,000 to 80,000 hectares annually, has resulted in tensions between these hunting communities and the forest products industry.[3]

The forest products industry entered this region approximately 35 years ago. By the mid-1970s, approximately 1.5 million m^3 of softwoods, mainly black spruce, were being extracted each year from an area between the 49th and 50th parallels.[4] The supply areas allotted to forestry companies totaled about 25,000 hectares.[5] Fifteen years later, after the abandonment of the supply guarantees and the introduction of the TSFMAs, that area had roughly doubled to 52,000 hectares. By 1995, the area allotted under the TSFMA tenure arrangement had expanded to 70,000 hectares, roughly three times the supply area 30 years ago.[6]

The TSFMA's represent long-term supply commmitments entered into by the Quebec government. They are based on mill capacity and cannot easily be scaled back without giving rise to claims for compensation. The result is considerable pressure on the available forest resources. Forests are now being logged with yields close to the current limit of viability of 50 m^3/ha (the average for this northern region is only 76 m^3/ha). In the process, the northern limit of viable forestry operations has extended northwards from the 50th parallel to the 52nd parallel and even above the 52nd parallel in some areas. Prospects for adequate regeneration are limited, and the constraints, both biological and economic, are increasingly evident.[7]

Yukon Timber Supply Analysis

The Yukon is home to some of Canada's remotest and slowest growing forests. Until recently, the territory's annual logging rate rarely topped 75,000 m^3.[8] But in 1993, the cut level began to rise in response to timber shortages in British Columbia and low Yukon stumpage fees. In 1997–98, the cut reached 386,000 m^3.[9]

The Yukon government's 1998 draft Timber Supply Analysis (TSA) estimates the yearly wood supply at 402,500 m^3 for the southern Yukon, where most of the merchantable timber is located.[10] However, 80 percent of this timber volume is from forests classified by the Yukon Forest Resources Department as "poor productivity sites" with trees 10 to 15 meters in height at 100 years of age. The department acknowledged that it is uncertain whether such marginal sites will adequately regenerate in the Yukon's cold, dry climate.

The draft TSA stipulates that harvestable spruce stands must have a minimum volume of 100 m^3/ha. Two independent reviews questioned these productivity and merchantability assumptions.[11] One review noted that including low productivity sites in the timber harvesting land base will have the effect of inflating the estimate of timber supply, with the result that industry will attempt to concentrate its AAC in more desirable stands.[12]

[1] L. Pederson, *Williams Lake Timber Supply Area, Rationale for allowable annual cut (AAC) determination* (Victoria: Ministry of Forests, 1996). See also: L. Pederson, *Williams Lake Timber Supply Area, Rationale for allowable annual cut (AAC) determination* (Victoria: Ministry of Forests, 1997). Online at: http://www.for.gov.bc.ca/tsb/tsr2/tsa/tsa37/ration/willadd.htm (February 3, 2000).
The Chief Forester revised the 1996 decision in 1997, but the AAC remains essentially unchanged.

[2] Ministère des Ressources naturelles du Québec, *Rapport sur l'État des Forêts Québécoises 1990-1994* (Québec: Gouvernement du Québec, 1996), pp. 111-113.
According to recent affidavits filed by the government the AAC is now up to 7.9 million hectares.

[3] Numbers here calculated by totaling all 5-year (1994-1999) TSFMA figures for the study area.

[4] Pierre Dugas, *Industrie Forestière: Rapport pour les Besoins du schéma d'amenagement* (Quebec: Société de la Baie James Environment et aménagement du Territoire, 1977), p. 14.

[5] Ministère de l'Énergie et des Ressources, Direction des Operations regionales, *COGEF Management Plans for Management Units 85, 86, 87 and 26* (Québec: Gouvernement du Québec, 1978-1979).
The COGEF plans were internal administrative documents prepared in several editions between 1978 and 1980 and were not published.

[6] Ministère des Ressources naturelles du Québec, *Rapport sur l'État des Forêts Québécoises 1990-1994* (Québec: Gouvernement du Québec, 1996).

[7] Ministère des Ressources naturelles du Québec, *Rapport sur l'État des Forêts Québécoises 1990-1994* (Québec: Gouvernement du Québec, 1996). See also: Ministère de l'Énergie et des Ressources, Direction des Operations regionales, *COGEF Plan d'Unité de Gestion de Chibougamau (no. 26)* (Québec: Gouvernement du Québec, 1978-1980), p. 58.

[8] DIAND Yukon Forest Resources, *Summation of Timber Harvesting, 1963 to Present.*

[9] DIAND Yukon Forest Resources, *1997/98 Timber Year Summary: Volumes Harvested.*

[10] DIAND Yukon Forest Resources, *Preliminary Timber Supply Analysis For the Southern Yukon, 1998.*

[11] Herb Hammond, Silva Ecosystem Consultants Ltd., *Review of Preliminary Timber Supply Analysis for the Southern Yukon* (Whitehorse: Teslin Tlingit Council and Yukon Conservation Society, 1998). See also: Douglas, H. Williams, Ph.D., Cortex Consultants, *Review of the Preliminary Timber Supply Analysis of the Southern Yukon* (Whitehorse: Department of Economic Development and Department of Renewable Resources of the Yukon Territorial Government, 1998).

[12] Douglas, H. Williams, Ph.D., Cortex Consultants, *Review of the Preliminary Timber Supply Analysis of the Southern Yukon* (Whitehorse: Department of Economic Development and Department of Renewable Resources of the Yukon Territorial Government, 1998), p. 7.

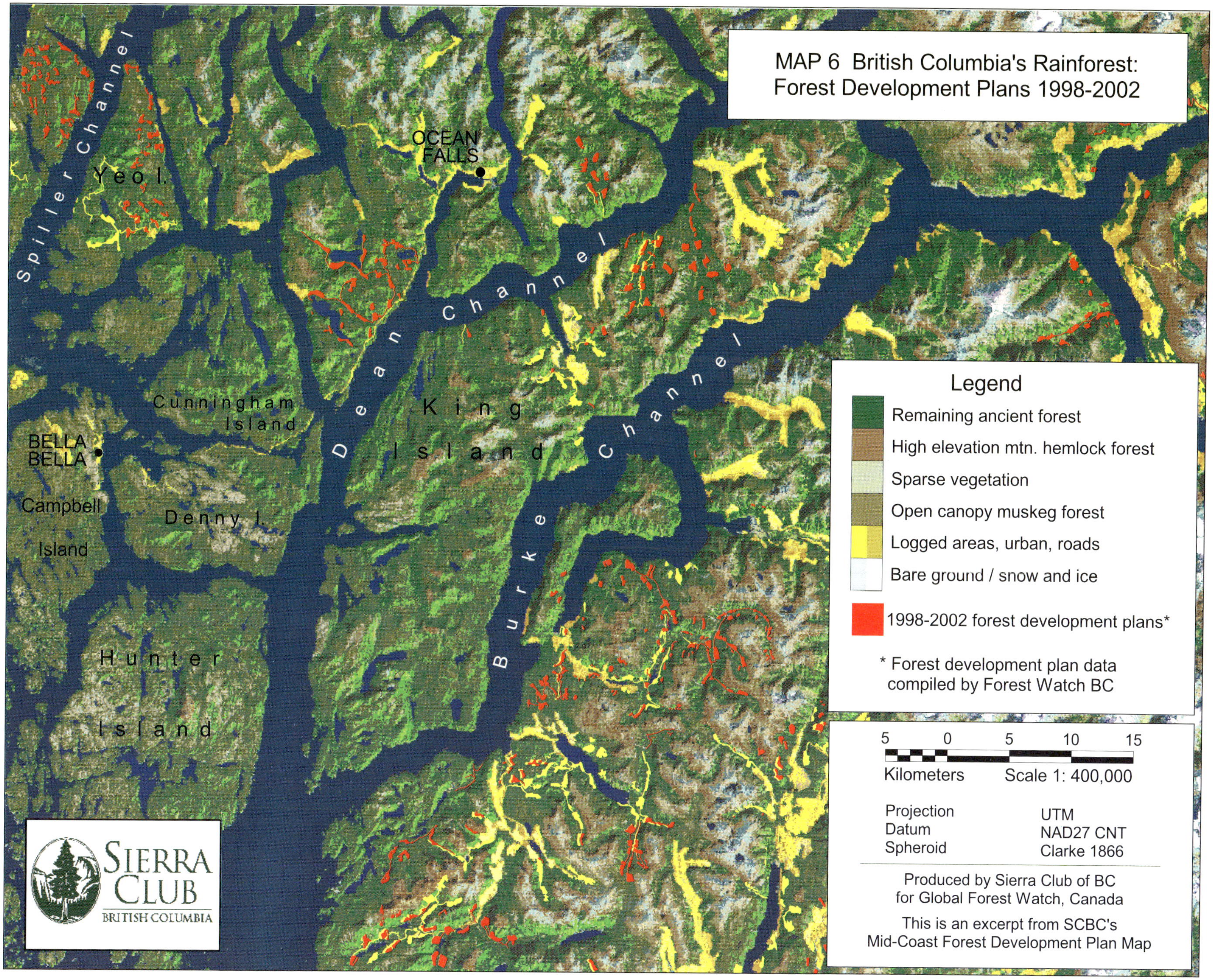
MAP 6 British Columbia's Rainforest:
Forest Development Plans 1998-2002
Spiller Channel
Yeo I.
OCEAN FALLS
Dean Channel
Burke Channel
Cunningham Island
King Island
BELLA BELLA
Campbell Island
Denny I.
Hunter Island
Legend
Remaining ancient forest
High elevation mtn. hemlock forest
Sparse vegetation
Open canopy muskeg forest
Logged areas, urban, roads
Bare ground / snow and ice
1998-2002 forest development plans*
* Forest development plan data compiled by Forest Watch BC
5 0 5 10 15
Kilometers
Scale 1: 400,000
Projection UTM
Datum NAD27 CNT
Spheroid Clarke 1866
Produced by Sierra Club of BC for Global Forest Watch, Canada
This is an excerpt from SCBC's Mid-Coast Forest Development Plan Map
SIERRA CLUB
BRITISH COLUMBIA

MAP 8 Operating Areas of Canada's Largest Forestry Companies

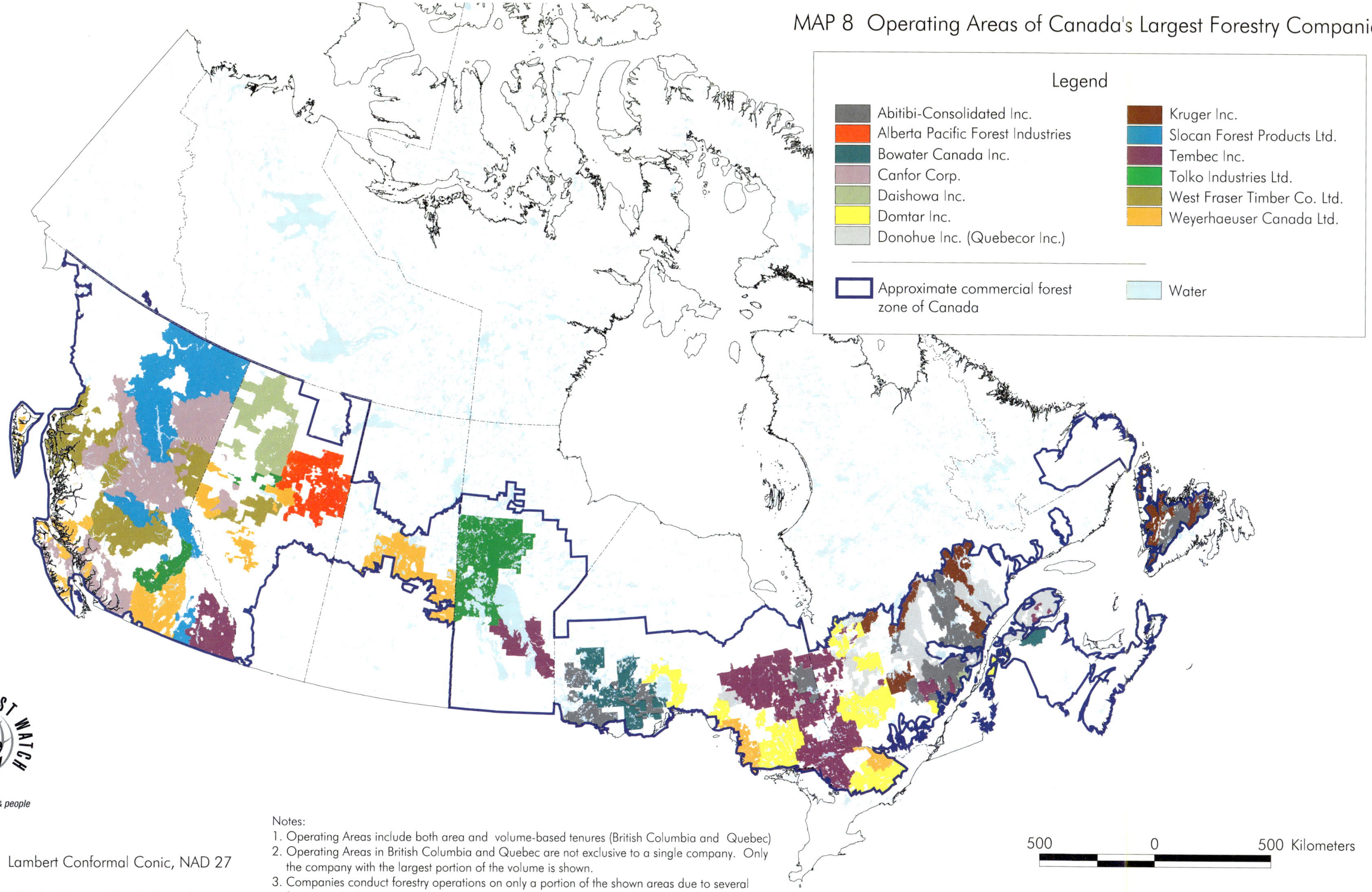

Scale: 1:20,000,000

GLOBAL FOREST WATCH

Linking forests & people

Projection: Lambert Conformal Conic, NAD 27

Produced by Global Forest Watch Canada

Notes:

1. Operating Areas include both area and volume-based tenures (British Columbia and Quebec)
2. Operating Areas in British Columbia and Quebec are not exclusive to a single company. Only the company with the largest portion of the volume is shown.
3. Companies conduct forestry operations on only a portion of the shown areas due to several factors, including operating constraints, terms of the tenure and government regulations.

MAP 7 Ecological Limitations to Commercial Forestry

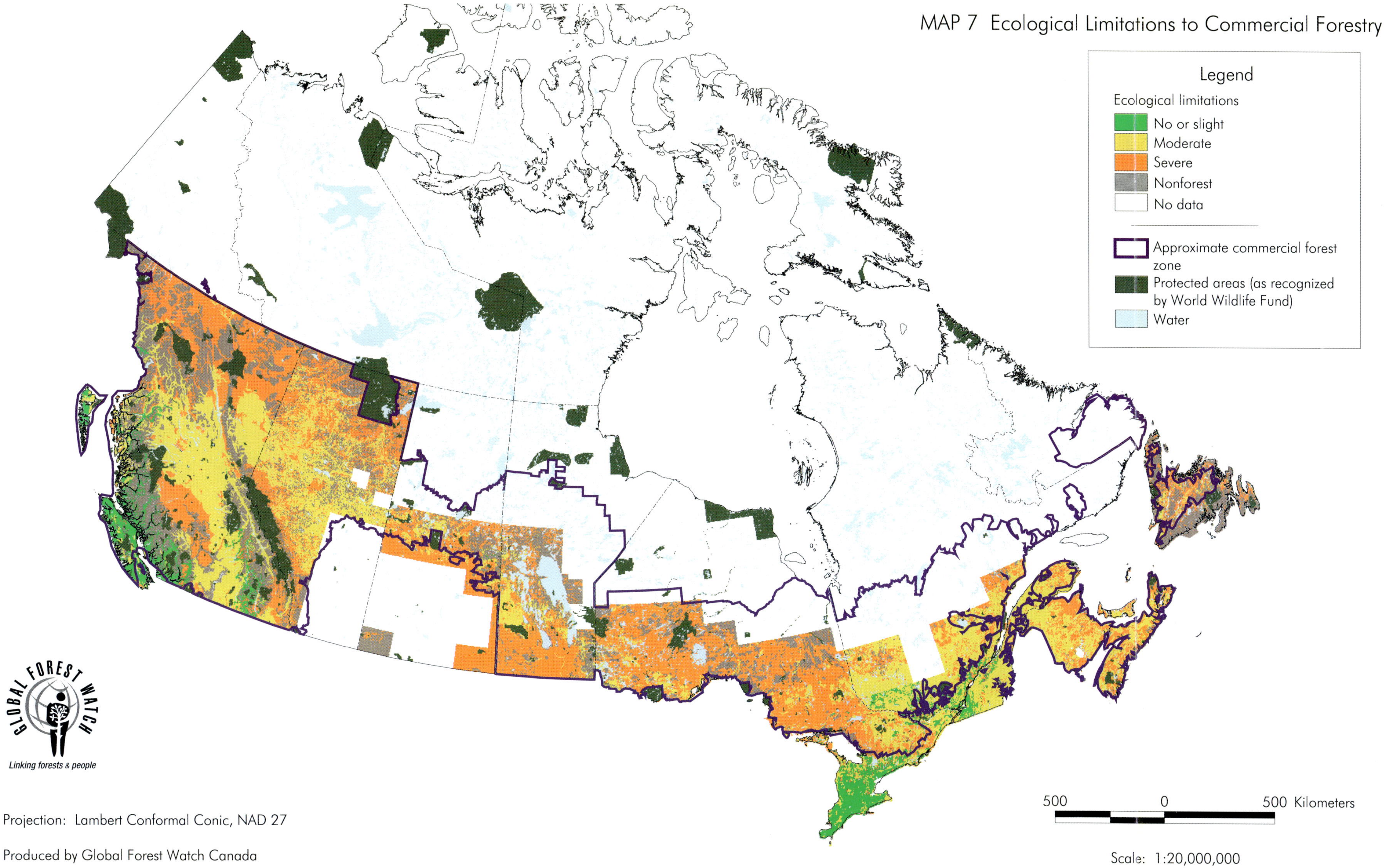

MAP 9 Corporate Concentration within British Columbia

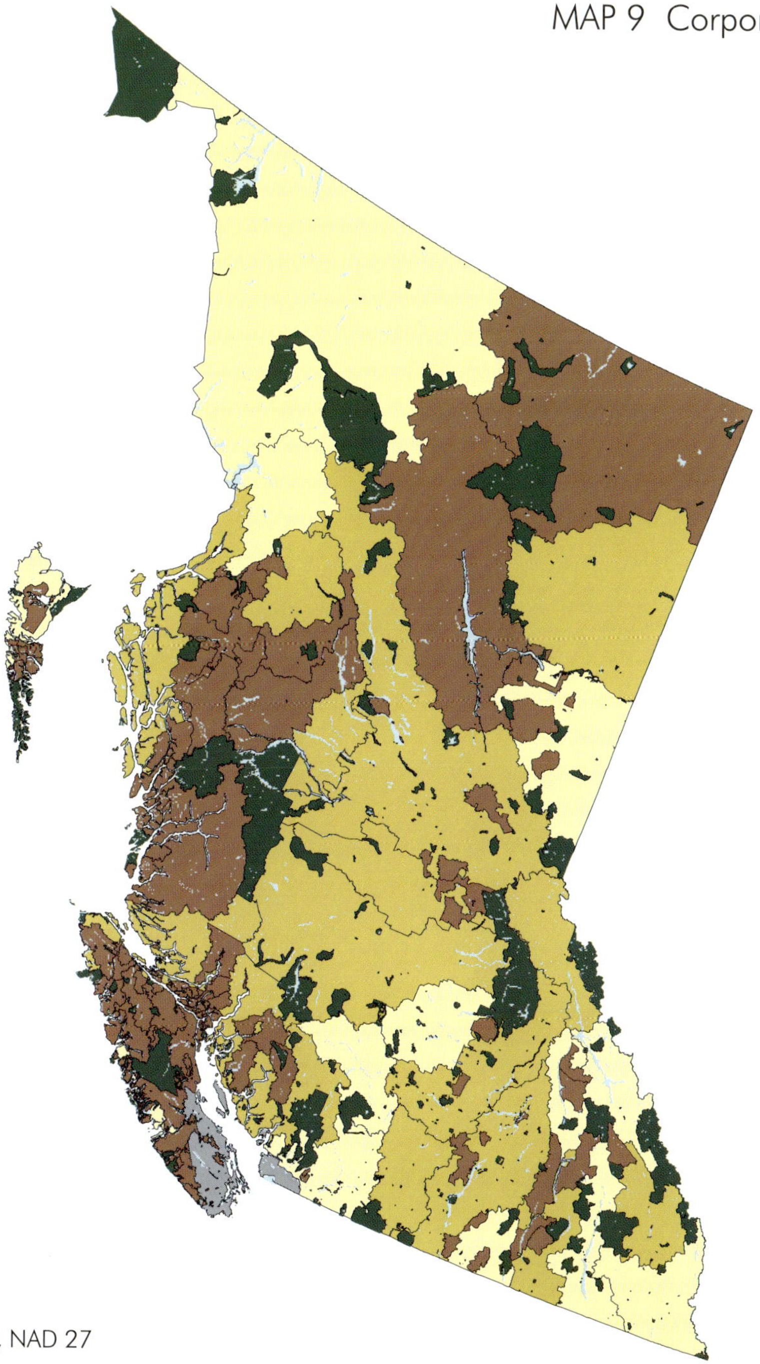

Legend

Percentage of allowable annual cut held by ten largest forestry companies (by administrative unit)

- 0 to 33%
- 33 to 66%
- 66 to 100%

- Private land
- Protected areas (as recognized by the World Wildlife Fund)
- Water

200 0 200 Kilometers

Scale: 1:10,000,000

GLOBAL FOREST WATCH

Linking forests & people

Projection: Lambert Conformal Conic, NAD 27

Produced by Global Forest Watch Canada

MAP 10 First Nation, Métis and Inuit Land Claims and Treaties

Inset: First Nation, Métis and Inuit communities and existing forests

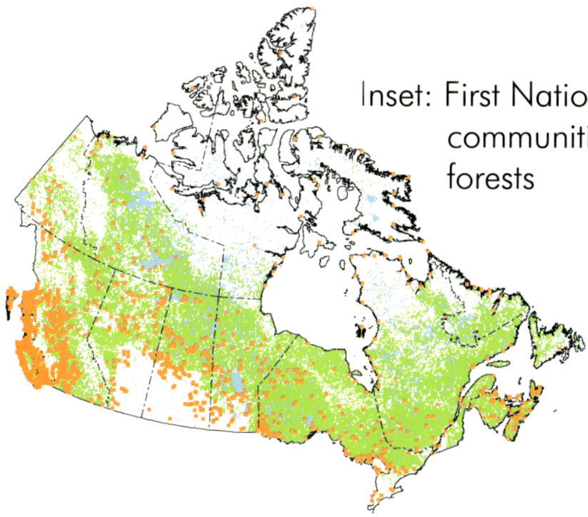

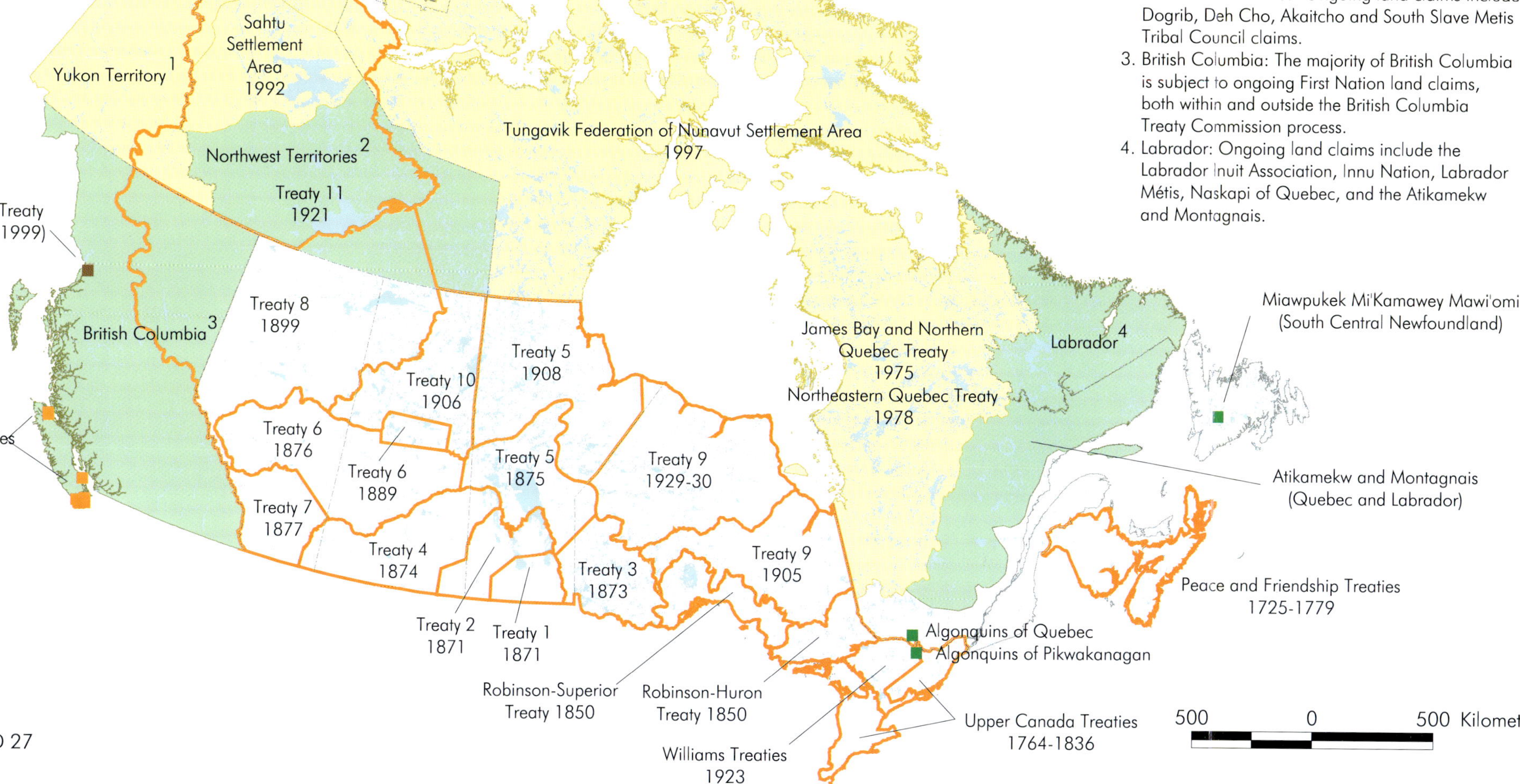

Legend

- Unsettled land claims
- Modern land claim settlements
- Historic treaties
- Water

Notes:

1. Yukon: The Umbrella Final Agreement includes fourteen First Nations, seven of which have settled their individual claims.
2. Northwest Territories: Ongoing land claims include Dogrib, Deh Cho, Akaitcho and South Slave Metis Tribal Council claims.
3. British Columbia: The majority of British Columbia is subject to ongoing First Nation land claims, both within and outside the British Columbia Treaty Commission process.
4. Labrador: Ongoing land claims include the Labrador Inuit Association, Innu Nation, Labrador Métis, Naskapi of Quebec, and the Atikamekw and Montagnais.

Linking forests & people

Projection: Lambert Conformal Conic, NAD 27

Produced by Global Forest Watch Canada

Scale: 1:25,000,000

Indicator 8: Regeneration

Most logged areas are left to regenerate naturally

Regeneration of forest lands after logging is an indicator of the sustained productivity of forest ecosystems. How forests are regenerated—whether naturally, or through reseeding—indicates the intensity of management. In general, more intensive management corresponds to higher timber yields, though it may also imply tradeoffs in terms of biodiversity values and other ecosystem services (particularly where regeneration does not mimic natural succession).

The federal government reports that approximately 60 percent of logged areas in Canada are left to regenerate naturally and 40 percent are seeded or planted, although the numbers vary significantly by jurisdiction.[81] For example, in British Columbia, the area prescribed for natural regeneration has declined from 50 percent in 1988 to less than 35 percent in 1992-1993.[82]

National data on the regeneration status of areas logged prior to 1974 are not available. (Some provincial jurisdictions do have data going back to the 1920s.) Regeneration efforts prior to 1975 were generally inadequate or nonexistent, and as a result there was a substantial backlog of "understocked" forests. (The term "understocked" refers to logged areas that require forest management treatments, such as planting or weeding, to meet established stocking standards for commercial tree species).

Total understocked forest area is declining

The backlog of understocked forests continued to increase throughout the 1980s, peaking at 2.7 million hectares in 1991.[83] Due to increasing reforestation efforts in the 1980s and 1990s, the total understocked area has begun to decline. As of 1996, the federal government reports that 2.5 million hectares of harvested Crown lands remain understocked.[84] As a result of more intensive planting and modified harvesting methods, annual area successfully regenerated to commercial species has increased by 23 percent since 1991.

Some recently logged areas are reported as understocked due to the time lag between logging and the results of silvicultural treatments or natural stand development. The federal government reports that since 1975 silviculture programs have ensured the successful regeneration of 90 percent of sites logged within 10 years of logging.[85]

For the 1975–1996 period, the amount of logged Crown forestland that is understocked varies widely by jurisdiction, from a low of 4 percent in Nova Scotia and New Brunswick to a high of 85 percent in the Northwest Territories.[86] *(See Figure 11.)*

Saskatchewan's situation is discussed in Box 7.

Figure 11. Understocked Harvested Crown Land, 1997

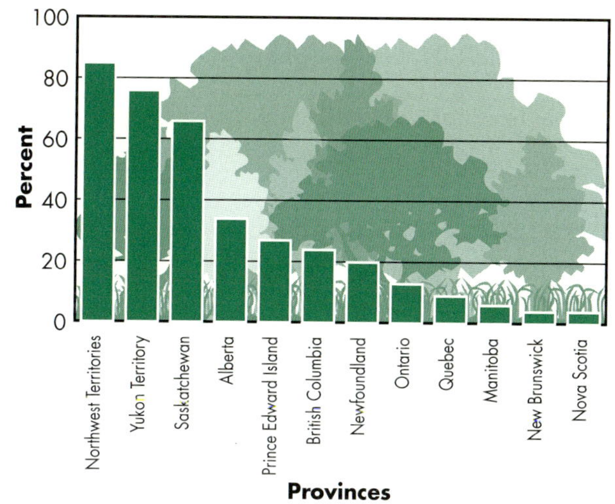

Source: Canadian Forest Service, *The State of Canada's Forests: 1998-99 Innovation* (Ottawa: Natural Resources Canada, 1999), pp. 26-32. Online at: http://nrcan.gc.ca/cfs/proj/ppiab/sof/common/latest.shtml (January 18, 2000).

BOX 7 Problems with Regeneration in Saskatchewan

Two thirds of forests logged in Saskatchewan since 1975 cannot be considered stocked, highlighting a serious forest management problem in the province.[1] Estimates on the total amount of Saskatchewan's backlog of not sufficiently re-stocked lands (NSR) vary widely, ranging from 263,000 hectares to 619,000 hectares.[2] These disparities have arisen because details about the backlog and inventory generally are thought to be incomplete, unreliable, or not up-to-date. This amount does not include the backlog prior to 1971, which has not been well documented.[3] Approximately 50 percent of Saskatchewan's understocked forest areas now lie on the northeast section of the province's commercial forest zone.[4]

The province is still responsible for all understocked land created as a result of fire and all understocked land created prior to the signing of current Forest Management Agreements.[5] Provincial forestry officials report that the strategy for tackling the backlog is simply "one tree at a time" and when the province can afford it.[6] There is not an active strategy for tackling the provincial backlog of NSR. In the meantime, on April 26, 1999, the government of Saskatchewan announced that it will take a number of measures to support private sector investment of over $850 million by Weyerhaeuser and other firms, which could create almost 10,000 new jobs in the sector within the next few years.[7]

[1] Canadian Forest Service, *The State of Canada's Forests: The People's Forests 1997-98.* (Ottawa: Natural Resources Canada, 1998), p. 29. Online at: http://nrcan.gc.ca/cfs/proj/ppiab/sof/sof98/toc.shtml (January 19, 2000). The data for the status of harvested Crown land reflect the cumulative area harvested since 1975. Data for private lands are not included. The term "stocked" refers to land where the forest cover meets certain timber-production standards established by forest management agencies in each province and territory. The term "understocked" refers to harvested land that requires forest management treatments, such as site preparation, planting, seedling or weeding, to meet established standards. This category also includes land that has not yet been surveyed. A signficant proportion of recently harvested areas will always be reported as understocked because of the time lag between harvesting and observable results of subsequent treatments. The small percentage of the area harvested each year that is devoted to access roads is not included in these data.

[2] Canadian Forest Service, *The State of Canada's Forests: 1998-99 Innovation* (Ottawa: Natural Resources Canada, 1999), p. 30. Online at: http://nrcan.gc.ca/cfs/proj/ppiab/sof/common/latest.shtml (January 28, 2000). See also: Delcan Western Ltd et al. *State of the Resource Report: Province of Saskatchewan Integrated Forest Resources Management Plan* (Prince Albert: Department of Environment and Resource Management and Forestry Canada, 1993), p.135.

[3] Delcan Western Ltd et al. *State of the Resource Report: Province of Saskatchewan Integrated Forest Resources Management Plan* (Prince Albert: Department of Environment and Resource Management and Forestry Canada, 1993), p.139.

[4] Jim Smith, Project Technician and Michael McLaughlan, Forest Ecologist, Saskatchewan Environment & Resource Management, Forest Ecosystems Branch, private communication, March 18, 1999.

[5] Jim Smith, Project Technician, Saskatchewan Environment & Resource Management, Forest Ecosystems Branch, private communication, March 18, 1999.

[6] Jim Smith, Project Technician, Saskatchewan Environment & Resource Management, Forest Ecosystems Branch, private communication, March 18, 1999.

[7] Government of Saskatchewan. April 26, 1999. "Major forest industry expansion announced: industry's plans could create up to 10,000 new jobs." Online at: http://www.gov.sk.ca/newsrel/1999Apr/347.99042602.html (January 28, 2000).

KEY ACTORS

Various stakeholders benefit directly or indirectly from forestry operations and have a vested interest in promoting sustainable timber production. These groups include forest companies, communities, aboriginal communities, the public, and consumers. The data on forest products described earlier provides some indication of how consumers benefit. Some of these groups are profiled below.

Indicator 9: Forestry Companies

While most forests are publicly owned, private companies do most of the logging. As discussed earlier, these companies obtain tenure or license rights from provincial and federal agencies.

10 companies generate over $24 billion in sales

In a survey of the global forest industry, 17 Canadian companies were in the Top 100 list for 1997. Ten Canadian companies accounted for $24.8 billion (US $17 billion) in sales in 1997. *(See Table 11.)*

13 companies hold tenures at least the size of Switzerland

The largest companies by area of operation are presented on *Map 8*. Each of the companies represented on this map have Canadian holdings larger than the area of Switzerland (41,000 km^2). Thirteen companies hold the equivalent of 48 percent of the area of commercial forest operations. *(See Table 12.)* This trend is indicative of a global trend towards consolidation in the forest products industry. Increasingly, responsible management falls on the shoulders of fewer, bigger companies.

10 companies hold almost 60 percent of the AAC in British Columbia

In two provinces—British Columbia and Quebec—a significant portion of logging rights are given out on a volume basis, or a total amount that can be harvested, rather than just on an area basis. These volume areas for British Columbia and Quebec are actually shared by a number of companies. To illustrate this, we mapped out areas of operation for 10 companies that harvest the largest volume of timber in British Columbia. *(See Map 9)*. Some 59 percent of ownership within Timber Supply Areas is held by these 10 largest companies. *(See Table 13.)* Similar analysis is needed for Quebec.

Table 11. Top 10 Canadian Forestry Companies by Sales, 1997

COMPANY	SALES ($U.S. MILLIONS)
MacMillan Bloedel (now Weyerhaeuser Canada Ltd.)	3,265
Abitibi Consolidated Inc.	2,706
Noranda Forest	1,641
Cascades	1,594
Avenor	1,438
Domtar Inc.	1,399
West Fraser Timber Co. Ltd.	1,350
Canfor Corp.	1,329
Donohue Inc. (Quebecor Inc.)	1,260
Fletcher Challenge Canada	975

Source: PriceWaterhouseCoopers, *Global Forest and Paper Industry – 1998 Edition* (Vancouver: PriceWaterhouse-Coopers, 1998).
Note: Abitibi Consolidated Inc. and Donohue Inc. merged on February 11, 2000. Avenor is now Bowater Canada Inc.

Table 12. Top Forestry Companies in Canada by Area of Operation

PARENT COMPANY	AREA OF OPERATION (km^2)
Slocan Forest Products Ltd.	132,325
Tembec Inc.	129,029
Weyerhaeuser Canada Ltd.	108,453
Tolko Industries Ltd.	106,324
Abitibi Consolidated Inc.	86,341
Domtar Inc.	83,784
Donohue Inc. (Quebecor Inc.)	79,011
Canfor Corp.	71,945
Daishowa Inc.	66,137
Alberta Pacific Forest Industries Inc.	58,772
West Fraser Timber Co. Ltd.	54,906
Bowater Canada Inc.	50,482
Kruger Inc.	48,083

Source: GFW Canada
Note: Only companies with operating areas larger than Switzerland (approximately 41,000 km^2) are shown.

Table 13. Top 10 Forestry Companies in British Columbia

COMPANY	TOTAL AAC (CUBIC METERS)	RANK IN BRITISH COLUMBIA AAC	PERCENT TOTAL OF BRITISH COLUMBIA
Canfor Corp.	8,305,438	1	11.8
Weyerhaeuser Co.	7,252,374	2	10.3
Slocan Forest Products Ltd.	6,209,038	3	8.8
West Fraser Timber Co. Ltd.	4,204,134	4	6.0
Doman Industries Ltd.	4,080,471	5	5.8
International Forest Products Ltd.	3,554,877	6	5.0
Skeena Cellulose Inc.	2,337,550	7	3.3
Riverside Forest Products Ltd.	2,306,776	8	3.3
Weldwood of Canada Ltd. (Champion International Corp.)	2,111,909	9	3.0
TimberWest Forest Corp.	1,492,596	10	2.1
Total (All Top Ten):	**41,855,163**		**59**

Source: Canadian Forest Service. *The State of Canada's Forests: 1998-99 Innovation* (Ottawa: Natural Resources Canada, 1999), pp. 26-32. Online at: http://nrcan.gc.ca/cfs/proj/pplab/sof/common/latest.shtml (January 18, 2000).

Indicator 10: Jobs and Wages

Over 1,600 communities depend on forest industry jobs

There are 337 forest-dependent communities, in which forest industries account for more than 50 percent of employment. A further 1,294 communities are classed as having some reliance on the forest industry.[87]

Employment is one indicator of the benefits communities derive from the forest industry. The forest industry remains a significant source of jobs and economic activity in many Canadian villages and towns. The industry generated over $11 billion (US $8.3 billion) in wages in 1996.

The Forest sector generates over 350,000 direct jobs

According to Statistics Canada data, the forest sector continues to contribute a significant number of jobs—about 350,000 direct jobs—despite fluctuations in various parts of the industry. *(See Figure 12.)*

Employment in wood industries—including sawmills, shingles and shake mills, veneer and plywood mills, wood boxes, and doors—grew 25 percent in the period between 1990 and 1998. Pulp and allied industry jobs have declined approximately 10 percent over the last 9 years.

Logging generated over $15 billion in wages and benefits in 1998

Forest industry jobs are comparatively well paid, with forest workers earning on average close to $61,717 (US $42,584) in 1997 ($47,179 in wages; $14,538 in benefits). This is substantially above national average earnings for that year of $38,011 (US $26,228).[88]

Figure 12. Jobs Per 1,000 m^3 of Harvested Timber, 1989-1997

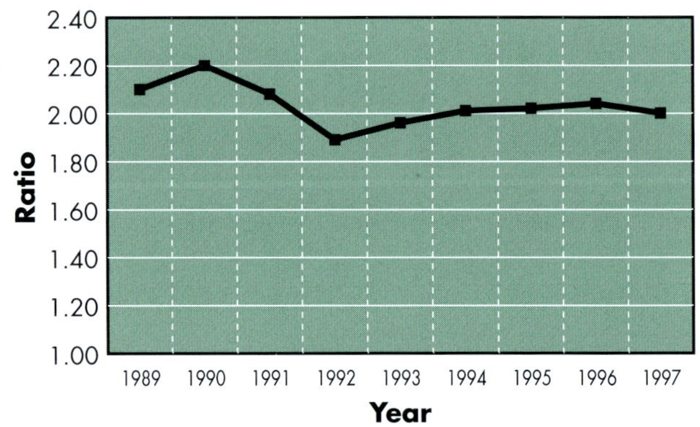

Sources: Canadian Council of Forest Ministers "National Forestry Database Program: Forest Products: Table 5.1." Online at: http://nfdp.ccfm.org (February 7, 2000). See also: Canadian Council of Forest Ministers. "National Forestry Database Program: Revenues and Economic Profile: Table 8.2." Online at: http://nfdp.ccfm.org (February 7, 2000).

Indicator 11: First Nations and Métis

80 percent of Canada's First Nations and Métis live within forested regions

Almost 80 percent of Canada's 1 million aboriginal people (First Nations, Inuit, and Métis) live on reserves and communities in boreal or temperate forests.[89] First Nations and Métis are mostly based in forest areas, while Inuit communities are located in mostly nonforested areas in the northern part of Canada. *(See Map 10.)* Further discussions in this section focus primarily on First Nations and Métis.

First Nations and Métis have extensive rights to forest lands

"Aboriginal and treaty rights" were recognized and affirmed by Canada's Constitution in 1982.[90] Given their historic presence in Canada, First Nations and Métis have certain rights that are still being deliberated and outlined in the legislative and court process. *(See Box 8.)*

Map 10 illustrates the location of aboriginal communities as well as the various categories of treaties and land claims across the country. Aboriginal rights and treaty and land claim processes have long-term implications for forests and forest management in Canada which could lead to widespread shifts in ownership and management of forest resources.

The government of Canada estimates that it has signed a number of treaties with First Nations.[91] These treaties cover most of Canada's forested regions. Many First Nations in British Columbia and throughout northern Canada never signed treaties with the federal government and are now actively negotiating comprehensive or specific land claims. The federal government has a comprehensive claims policy in place.

In British Columbia, where treaties were never signed, the provincial, federal, and First Nations Summit established the British Columbia Treaty process in 1992. Sixty percent of First Nations are involved in the treaty process, which is designed to address issues related to aboriginal rights and title. Approximately 300 First Nations are now involved in 80 negotiations involving some aspect of self-government. These settlements or modern treaties usually include ownership of a specific land base, wildlife harvesting rights, participation in management decisions, financial compensation, and resource revenue sharing.[92]

To date, several agreements have been reached in Canada's north, including the Inuvialuit Final Claim, The Nunavut Final Agreement (1992), Gwich'in Final Agreement (1992), Sahtu Dene and Métis Final Agreement (1993), the Yukon Comprehensive Land Claim Agreement, and six specific claims. In the Yukon, for example, six First Nations recently won jurisdiction over 41,000 square kilometers of forest. The first modern treaty in British Columbia, the Nisga'a Final Agreement, was signed in 1999 but is not yet ratified. Still, there are nearly 100 outstanding claims in British Columbia alone.[93]

Although some First Nations will likely settle for monetary compensation, many of the First Nations that reside in forests will claim forested land.[94] First Nations are growing increasingly frustrated with the pace of land claim settlements.[95] Several First Nations are now logging Crown lands without government authorization in British Columbia.[96]

First Nations reserves in southern Canada encompass less than 1 percent of all productive forest land

First Nations reserves account for 3 million hectares, of which 1.4 million hectares (0.6 percent of Canada's productive forests) are suitable for sustainable resource uses such as logging.[97] In Alberta, there are also some Métis communities that have settlement lands. As noted above, a number of First Nations in northern Canada have signed land claim agreements that provide them private lands as well as a role in natural resource management in the entire claim area.

First Nations have some direct tenures

First Nations have direct forest tenures in some jurisdictions and are participants in a number of joint ventures. Limited data is available on First Nations tenures (timber leases) in Canada. In British Columbia, First Nations have approximately 4 percent of all commercial rights, the most significant portion in the country. Other jurisdictions, such as Alberta, do not track tenures owned by First Nations.[98] *(See Table 14.)*

Many provinces, including British Columbia, Alberta and Saskatchewan, are now negotiating with First Nations about timber cutting rights. A recent British Columbia study recommended that First Nations with ecologically sound logging plans "should be given priority for any new allocation of industrial forestry tenures."[99] Recent court rulings and political developments, however, have not yet been thoroughly addressed and First Nations involvement in forestry has not yet been assessed in terms of current official wood supply estimates."[100]

A number of government and industry initiatives have recognized the right of First Nations to be more involved in the forest sector.[101] These programs have included business loans, training, federal programs (First Nations Forestry Program), and joint venture agreements with industry. A 1994 survey of 15 companies found that most companies believe that shared management and greater participation of First Nations and Métis in forest management decisionmaking is likely to increase.[102]

Roughly 2 percent of First Nation members and Métis are employed by the forest sector.

First Nations members and Métis are slowly gaining more benefits from the forest sector. In 1991, the forest sector employed 2.2 percent of the aboriginal work force, or 10,100 people.[103] In 1998, under 20 First Nations members worked as foresters.[104] First Nations and Métis businesses are also involved in the primary resource sector including forestry; 17 percent of 20,000 companies are in the forest sector.[105]

Land Use Conflicts

Commercial forestry and other resource developments can also conflict with traditional First Nations values towards the land. The Royal Commission on Aboriginal Peoples documented widespread alarm about the scale and pace of industrial developments. First Nations and Métis communities have become increasingly "alienated from their traditional dependence on forest lands." [106]

Table 14. First Nations' Tenures for Selected Jurisdictions

PROVINCE	DETAILS
Alberta[1]	3 cooperative management agreements in place with First Nations. The Alberta Government does not track First Nations ownership of tenures or forest permits.
British Columbia[2]	Tenures awarded by volume. Tenures vary widely by length and amount. Some are for given volumes per year for periods ranging from 3 to 15 years. Others are for a set volume for a specific time period. Total AAC to First Nations is 852,328 m^3/year. (This amount was calculated by averaging the amount per year from each 100 percent First Nation owned tenure.)
New Brunswick[3]	In 1997, the province decided to proportionally allocate 5 per cent of the total allowable cut on Crown Land (public land) to the 15 First Nations, based on population size. Each First Nation was asked to sign a one- or multi-year agreement without prejudice to any Native land claims. The volume allocated totals 185,000 m^3 of softwood and 50,000 m^3 of hardwood. First Nations also receive royalties on this volume (approximate value of $2.6 million dollars). Twelve out of 15 agreements were signed in 1999.
Newfoundland[4]	Seventy percent of the province is tied up between two companies, neither of them First Nation. No major licenses to First Nations.
Saskatchewan[5]	In April 1999, the Saskatchewan government announced major expansion in forest industry in northern Saskatchewan that will allocate timber for community-based partnerships. An example of existing First Nations involvement in forestry is the joint venture that the Meadow Lake Tribal Council initially held with Norsask Forest Products (which is now fully owned by the Tribal Council). There is also a Forest Management Licence Agreement (FMLA) held by Norsask and comanagement structures created by Mistik, the forest management company created by Millar Western and Norsask to manage the FMLA.

Note: This table is not intended to be inclusive of all tenures or joint ventures involving First Nations.

Sources:

1 Jamie McNeil-Honda, Negotiator and Policy Advisor, Alberta Environment, private communication, September 24, 1999.

2 Bill Dexter, Senior Program Advisor, Aboriginal Affairs Branch, British Columbia Ministry of Forests, private communication, September 23, 1999. "Tenures Awarded to First Nations by Forest Region" (01/26/99). The figures from the table obtained from Aboriginal Affairs Branch presented here only represents full tenures awarded to First Nations for Forest Licenses and not woodlots or tree farm licenses. British Columbia also tracks First Nation tenures that are joint ventures and tenures involving benefit agreements.

3 Julius Tarjan, Government of New Brunswick, private communication, October 18, 1999.

4 Darryl Harris, Newfoundland Government, Cornerbrook, private communication, September 24, 1999.

5 Val Nicholson, Saskatchewan Environment and Resource Management, private communication, September 27, 1999. Monique Ross, private communication, September 28, 1999. See also: News Release "Major Forest Industry Expansion Announced: Industry's Plans Could Create Up To 10,000 New Jobs" (Regina: Government of Saskatchewan, April 26, 1999). Online at: http://www.gov.sk.ca/newsrel/1999Apr/347.99042602.html (September 28, 1999).

BOX 8 Legal Issues Relating to First Nations' Rights

Several legal cases on treaty interpretation, implementation, and enforcement could alter the management and ownership of forests in many parts of Canada:[1]

- In northern Quebec, the Grand Council of the Cree have started court action against the governments of Canada and Quebec for breaching conditions of the 1975 James Bay and Northern Quebec Agreement. The Cree contend that the doubling of clearcut logging on their land in the last 20 years is a violation of international, aboriginal, and treaty rights.[2]

- In the Atlantic provinces, the Mi'kmaq are claiming the right to harvest timber on Crown land on the basis of treaties signed with the British government in the 1760s. This dispute, now in the courts, began last year with the arrest of 31 people for illegal logging.[3]

- On the prairies, the First Nations of Treaty 7 and 8 are contesting the legality of the Natural Resources Transfer Act, which transferred lands, mines, minerals, and royalty rights from Canada to three prairie provincial governments in 1930. First Nations elders argue that the treaties agreed that they would share resources such as timber; they did not give the rights to resources up altogether.[4]

A series of court decisions have addressed First Nations claims to Canada's natural resources, including forests. These controversial rulings may change provincial forest tenures as well as the way forest companies manage their operations on Crown land. They have also created widespread economic uncertainty and political confusion about the future of Canada's forest industry.[5]

- **The Sparrow decision (1990):** In the allocation of resources (such as fish and wildlife), federal and provincial governments must give aboriginal rights first priority after provision for conservation (or provide fair compensation).[6]

- **The Delgamuukw decision (1997):** The oral histories of aboriginal people can establish title to land, meaning ownership of the land and its resources. This title includes both traditional and nontraditional activities on issues of resource management.[7]

Court cases and legal decisions will continue to affect the management of resources for years to come.

[1] R. David House. 1998. "Aboriginal claims and the forestry industry: Claims processes and recent developments in the courts." *The Forestry Chronicle* 74(3): 334-341.

[2] Grand Council of the Cree. July 15, 1998. "Quebec Forestry Practices Violate Cree Rights." Online at: http://gcc.ca/News/quebec_forestry_practices_violat.htm (February 2, 2000).

[3] Graeme Hamilton, "Mi'kmaq to show they never gave up land," *The National Post* (July 20, 1999): A5.

[4] Bob Brown, "Natural Resources Transfer Act-You be the Judge," *Aboriginal Times* (April 3, 1999): 17-19. See also Tom Flanagan, "The Marshall ruling puts Western Canada's Economy in jeopardy," *The Globe and Mail* (October 7, 1999): A19.

[5] John Snow Jr., "Law of the Land," *Aboriginal Times* (April 3, 1999): 9-16. See also: R. David House. 1998. "Aboriginal claims and the forestry industry: Claims processes and recent developments in the courts." *The Forestry Chronicle* 74(3): 334-341.

[6] John Graham and Ioanna Sahas Martin. 1999. "Exploring the relationship between Aboriginal peoples and the Canadian forest industry." *Forestry Chronicle* 75(1): 67-72.

[7] R. David House. 1998. "Aboriginal claims and the forestry industry: Claims processes and recent developments in the courts." *The Forestry Chronicle* 74(3): 334-341.

SECTION 4. COMMITMENTS AND LEGISLATION

INTRODUCTION

As noted in sections 2 and 3, Canadians value their forests for a wide range of uses. They expect forests to be managed so that valued goods and services are provided over the long term. This section describes existing Canadian commitments and legislation regarding the sustainable management of forests. It responds to GFW's mandate to 1) monitor the implementation of laws and regulations established in the interest of forest stewardship and 2) identify the actors—including companies, individuals, government agencies, and others—engaged in development.

In this section, we set out to answer the following questions:

- What government legislation and commitments are in place to support forest stewardship?
- To what extent have these laws and commitments been implemented?
- What types of voluntary initiatives are under way?

We focus specifically on international commitments made by the federal government; federal and provincial legislation; national and provincial initiatives and commitments; and voluntary initiatives.

It is beyond the current scope of GFW Canada to assess the adequacy of existing legislation. We limit ourselves to reporting on existing key commitments, legislation, and initiatives.

Summary of Progress in Commitments and Legislation

In recent years, Canada has put in place a host of new, often high-profile initiatives and policies to promote forest stewardship. These include signing many international agreements; launching a National Forest Strategy and Forest Accord; initiating a criteria and indicators process to measure progress towards sustainable forest management; and developing a model forest network to apply new management approaches in 11 forests. Canada also has expanded its system of protected areas to now include 7.6 percent of the nation's forest area, though many natural regions are still not represented in the current system. These efforts indicate an increasing commitment to manage forests for a range of environmental goods and services rather than simply for timber production.

In this section, we list examples of such initiatives. We also provide partial information—largely derived from independent reviews commissioned by various government agencies—on progress in implementing new policies and agreements. In no way do these results constitute a comprehensive review of compliance. Rather, they highlight possible trends that may warrant further monitoring.

According to several study panels, there has been relatively little progress in implementing key international agreements such as the Convention on Biological Diversity, which could help protect forest species and ecosystems. At both the federal and provincial levels, deep budget cuts have drawn down the staffing and resources required to implement and enforce new policies and legislation. In response to these cuts, several panels have said that Canada's current federal forestry effort is in jeopardy.

Independent audits by Global Forest Watch Canada partners in Ontario and British Colombia identified numerous violations of existing management norms, which may be linked to poor enforcement. Increasingly, government agencies are transferring oversight responsibilities to industry, in the process potentially abdicating their responsibilities and obligations to First Nations. Several voluntary initiatives are under way to promote better harvesting practices, but data from Alberta and British Colombia indicate that voluntary efforts are less effective than active oversight by relevant government agencies, which are charged with managing 94 percent of Canada's forests.

Because of the difficulty of systematically assessing these areas, we have focused on short descriptions of each issue and examples of progress and problems with implementation. We have relied, to a large degree, on assessments carried out by major agencies and panels in Canada, including the Commissioner of the Environment and Sustainable Development, the Federal Standing Senate Committee on Agriculture and Forestry, the House of Commons Standing Committee on the Environment and Sustainable Development, and the National Round Table on the Environment and the Economy. Audits by environmental groups are also referenced.

INTERNATIONAL COMMITMENTS

Canada has signed over 230 international environmental agreements

The Canadian Government has now signed more than 230 international environmental agreements.[107] These agreements include commitments to protect forests around the Great Lakes, reduce greenhouse gases, arrest transboundary air pollution, and preserve wetlands. Canada has made several international commitments to protect its forests and was one of the first signatories of the UN Convention on Biological Diversity at the Earth Summit in 1992. This agreement obliges Canada to protect its biological resources and respect aboriginal rights to use those resources.

Canada also participates in international sustainable forest management initiatives, including the Montreal Process, which in 1995 committed parties to the tracking of progress toward sustainable management through a series of common criteria and indicators.[108]

Canada actively promotes international forest product trade liberalization initiatives, such as the recent round of negotiations at the World Trade Organization meeting in Seattle, Washington; the Asia-Pacific Economic Cooperation (APEC); and bilateral trade accords. Canada is also an active promoter of an international forest convention[109] and was a key player in the initiation of the Costa Rica Initiative to build global support for a forest convention.[110]

Implementation of global commitments is weak

Brian Emmett, Canada's Commissioner of the Environment and Sustainable Development, has examined Canada's progress in implementing international agreements.

In a 1998 report to the House of Commons, Emmett reported that "far too often" the government of Canada was failing to meet the promises made to Canadians and the international community in environmental matters; that many of these failures could be traced to poor management; and that the government had a great deal of work to do to handle the environmental challenges of the 21st century.[111]

His office examined two key international agreements, the United Nations Framework Convention on Climate Change and the United Nations Convention on Biological Diversity, both signed at the Earth Summit in Rio in 1992. His findings were:

- *Canada will not meet its commitment to curb greenhouse gas emissions.* Under the Climate Change Convention, Canada agreed to reduce greenhouse gas emissions. Canada's national response to the Convention has been to develop the National Action Program on Climate Change. Initially, Canada had committed to stabilize greenhouse gas emissions at 1990 levels by the year 2000. The Commissioner found that there was no implementation plan and that federal, provincial, and territorial roles have not been clearly defined. Initial assessments indicate that Canada will not be able to keep this promise, and that it will in fact exceed its stabilization target for the year 2000 by at least 11 percent.[112]

- *Canada has no overall strategy for implementing its Biological Diversity Strategy.* Under the Convention on Biological Diversity, Canada committed to creating and implementing a biodiversity strategy. Canada's national strategy was initially drafted in 1994, was released in November 1995, and received ministerial endorsement in April 1996. The Commissioner on the Environment and Sustainable Development reported that it was too early to undertake a complete assessment of implementation. He did state, however, that Canada was behind in its commitments. He concluded that better management was needed; a federal implementation plan was essential; and future reporting must reflect progress against measurable targets.[113]

Other reviews have noted that a significant part of the problem with implementation of international agreements, is that the federal government signs agreement but responsibility for implementation often rests with provinces.[114] Implementation by the federal government continues on both of these agreements. Updates on climate change can be found at http://www.ec.gc.ca/cc/CoP5/index_e.htm and on biodiversity at http://www.bco.ec.gc.ca/index_e.htm.

NATIONAL AND PROVINCIAL LEGISLATION

There are over 15 forest management laws in Canada

The Canadian Constitution gives authority over natural resources, including forests, to the provinces. Because of the complex ownership of forests in Canada, a number of laws are in place to oversee Canada's forests. (These are listed and described on our web-site at http://www.globalforestwatch.org). The provincial laws are the primary laws for forest management. Legislation increasingly reflects the public's interest in managing forests for a variety of uses rather than simply for timber supplies. New laws have been passed in recent years in a number of provinces, including British Columbia, Ontario, and Saskatchewan.

A number of environmental laws affect forests

A number of federal environmental laws also affect activities in forested areas.[115] These include, but are not limited to:

- The Fisheries Act, which provides for the protection of fish and fish habitat. This law's provisions must be considered when assessing major development activities, including activities in forested areas.

- The Canadian Environmental Protection Act, which provides for environmental protection and pulp and paper mill regulations.

- The Canadian Environmental Assessment Act, which requires assessments of federally funded and managed projects as well as projects on federal lands.

There are also a wide range of provincial and territorial environmental and resource laws that affect forests and how they are managed.

There has been no systematic evaluation of Canada's environmental and forestry laws and their implementation. This section provides examples of some reported implementation problems.

The national forestry budget has been reduced by 58 percent

The Canadian Forest Service (CFS), now a sector of the federal department of Natural Resources Canada, has been in existence since 1899. In recent years, CFS has experienced dramatic reductions in budget and staff. From 1995 to 1998, the annual operating budget of the CFS declined from $219 million to $93 million.[116]

Several reports have been critical of these changes. The Blue Ribbon Panel reports that many senior forestry officials now believe that "Canada has fallen well below a minimum safe level of core competence in all aspects of forestry research including research on forest ecosystems."[117] In 1996, the Canadian Institute of Forestry, a society that represents the nation's professional foresters, issued a statement that a strong federal presence should be maintained and that Canada's forests are too vital "to allow any further reduction in the size and effectiveness of Canada's federal forestry effort."[118]

Federal environmental enforcement capacity is hampered by budget cuts

Budget and staffing cuts have weakened the application of federal environmental laws. The House Standing Committee on Environment and Sustainable Development has noted that Environment Canada, the federal department with responsibilities for environmental protection, had its budget reduced by 40 percent and was actively moving toward voluntary compliance with environmental laws as a result. Their key findings include:

- There are only 8 full-time federal enforcement officers in Quebec and only 15 inspectors in northern and western Canada.[119]

- Polluters routinely escape prosecution or conviction because of government paperwork mistakes.[120]

- In 1996, 20 pulp and paper mills in Quebec discharged toxic effluents above legal standards without being prosecuted.[121]

- Environment Canada does not have comprehensive, standardized, and readily accessible data on enforcement budgets and expenditures.[122]

- The Fisheries Act is not enforced in some parts of the country because of lack of staff or jurisdictional disputes.[123]

- When provincial governments fail to uphold federal laws under bilateral agreements—a process known as harmonization—"the federal government has not intervened and taken enforcement action."[124]

Enforcement has been shown to raise compliance levels

Data compiled by Environment Canada on British Columbia's pulp mills illustrate the weakness of solely relying on voluntary compliance. From 1983 to 1991, the province's pulp mills and other industries tried to reduce the release of toxic wastes through a voluntary program. With voluntary compliance, most industries only achieved an average 60 percent implementation rate of best management practices. As a consequence, the salmon-bearing Fraser River still experienced chronic pollution. After Environment Canada targeted six mills for investigation and prosecution under the Fisheries Act, the discharge of toxic effluent decreased by over 90 percent and compliance rates climbed to 94 percent. These studies suggest that a strong enforcement program is necessary to provide incentives for voluntary initiatives.[125]

Ontario has reduced forestry staff and enforcement

As discussed earlier, every province has its own set of forestry laws, codes, and systems for compliance auditing. GFW Canada has focused on two key forestry provinces, Ontario and British Columbia, and examined compliance and enforcement issues.

In Ontario,

- In the last 4 years, more than 40 percent of staff at the Ministry of Natural Resources (MNR)—the department responsible for forests—has been laid off. Staff cuts particularly affected forestry branches dealing with compliance, monitoring, and policy.[126]

- Funding for forest management activities fell by $45.9 million in 1997–98.[127]

- Responsibilities for planning, inventories, monitoring, and silviculture have been transferred to the forest industry. The ministry now relies on company reports as its primary source of information on the state of the province's forests.[128]

Concerns are expressed in a number of recent reports:

- An audit of compliance with forestry guidelines in the Algoma Highlands in northern Ontario found that logging activities had threatened waterways, damaged fish-bearing streams, and left piles of garbage in remote areas. Violations were found in 55 percent of the sensitive sites designated for protection as Areas of Concern and Riparian Reserves.[129] An independent MNR field investigation later confirmed 10 specific contraventions and recommended enforcement action in three cases.[130]

• In 1998, the Ontario Divisional Court declared the Elk Lake, Upper Spanish, and Temagami forest management plans to be "of no force and effect." The court found that the Ministry of Natural Resources had failed to comply with the Crown Forest Sustainability Act with respect to work schedules, sustainability indicators, and timetable extensions.[131]

• The Environmental Commissioner of Ontario "raised concerns about forestry monitoring and compliance" in annual reports for 1996 and 1998.[132]

British Columbia's enforcement practices require further improvement

In June 1995, British Columbia enacted the Forest Practices Code.[133] The Ministry of Forests and the Ministry of Environment, Lands, and Parks acknowledged that implementation of the code has not been completed. Environmental nongovernmental organization (ENGO) audits of British Columbia's forest practices, based on government data, highlight the following:

• In 1996, 83 percent of forest streams were being clearcut to both banks, a potentially damaging practice permitted under the code for many streams.

• Approximately 40 percent of streams were not identified or were misclassified on forestry plans.[134]

• Fish streams in U.S. National Forests in the Pacific Northwest receive better protection than streams in British Columbia under the code.[135]

• Although 54 months have passed since the code was enacted into law, none of the three promised on-the-ground mechanisms intended to protect wildlife and biodiversity in British Columbia have been implemented.[136]

• British Columbia government documents note that biodiversity conservation is in jeopardy and recent staffing and budget cuts make it difficult to enforce the code.[137]

• Forest Development Plans that propose clearcutting on landslide-prone terrain are approved in violation of the code.[138]

• From 1995 to 1996, the British Columbia government laid no charges under the code.[139]

The Ministry of Forests and the Ministry of Environment, Lands, and Parks did a follow-up review of the ENGO report on stream-side forestry practices. This review looked at a subset of the streams audited by the ENGO and determined that 30 percent of streams were misclassified and 14 percent were not identified at all.[140] The review team's finding largely substantiates the ENGO audit conclusions; small differences are explained by the review team's smaller sample size. As a result of the problems it documented, the review team made 10 recommendations for improvements in stream-side management; however, few if any of the recommendations have been implemented in the ensuing 34 months.

In mid-1998, the Forest Practices Board conducted a special investigation of stream-side logging practices in British Columbia. The board found that 39 percent of all streams were misclassified and more than 50 percent of the smallest fish streams were misclassified.[141] Mirroring the findings of the Sierra Legal Defence Fund's (SLDF) audit, the board concluded that "the identification and classification of smaller fish streams (less than five meters wide) was considered the most significant problem area encountered during the investigation."[142] The board found that less than 5 percent of timber remains along the banks of approximately 20 percent of fish streams and 56 percent of non fish-bearing streams.[143] Neither the joint ministry review nor the board's special investigation specifically addressed the primary issue raised by SLDF, that of the total percentage of streams clearcut to their banks.

Box 9 looks at cases of noncompliance under the code.

INITIATIVES

In Canada in the last 10 years, there have been a number of forestry initiatives involving government, industry, First Nations and Métis, environmental groups, and others. A few of the key initiatives are reviewed below.

BOX 9 The British Columbia Forest Practices Code

Examining Enforcement of Forestry Law

In theory, British Columbia has one of Canada's most sophisticated legal regimes designed to ensure sustainable forestry. The law is known as the Forest Practices Code of British Columbia Act. When the code came into effect in 1995, the provincial government promised "tough enforcement."[1]

The Ministry of Forests has assumed the lead role in enforcement of the code.[2] The ministry's main objective is "to help the British Columbia forest industry carry out sound forest practices." This has led the ministry to focus its efforts on ensuring compliance, primarily through its inspection regime.[3] According to the ministry, "enforcement, by itself, is an inefficient and ineffective way to protect British Columbia's forests."[4]

For the past four years, the forest practices board, an independent government body, has been monitoring and reporting on compliance with and enforcement of the code. Despite significant improvements in forestry practices since enactment of the code, the board found there was "a need for better compliance with code requirements."[5]

The board raises questions about the quality of the inspections carried out by Ministry of Forests staff and subsequent enforcement actions. During an investigation of compliance and enforcement activities at Homesite Creek, the board found that over the course of six days, Ministry of Forests staff had conducted 1 pre-operational meeting and 12 logging inspections, yet the ministry failed to note the fact that logging was seasonally prohibited.[6]

In 16 random audits completed by May 1999, the board identified a total of 19 cases of "significant noncompliance" with the forest practices requirements of the code. These cases placed environmental values at risk in six forest districts.[7] The cases included inadequate protection of streams and fish habitat; poor road building and maintenance; and failure to comply with government-approved land use plans.[8] In 13 of the 19 cases, Forest Watch of British Columbia found that the Ministry of Forests failed to identify the significant noncompliance prior to the board's random audit. In every instance, after the Ministry of Forests had been informed of the board's finding of significant noncompliance with the code, they had not taken any enforcement action.[9]

The code allows for fines of up to $1 million a day to penalize poor forestry practices. Despite this, there has been a reported total of $2.3 million in tickets and penalties collected under the code between June 15, 1995 and July 14, 1998.[10] In comparison, during this same time period, the Vancouver Public Library collected $3.5 million in library fines.[11] In the case of a recent landslide, a penalty of $7,500 was issued for failure to adequately construct a logging road.[12] This penalty is the largest penalty that has been issued under the code in the Arrow Forest District. However, the costs of repair to the highway damaged by the slide, excluding the cost of damage to the soil, trees, transmission lines and lake, was $73,000.[13]

The Ministry of Forests' accounting practices create difficulties in generating accurate figures for program activity expenditures on enforcement.[14] Since the code took effect, the reported total expenditures on enforcement by Ministry of Forests staff at the regional and district level has been $25,000.[15] While the reported budget may not accurately reflect the true level of program activity, there have been only two prosecutions of a major forestry company in this time period.[16]

The Forest Crimes Unit of the Royal Canadian Mounted Police has reported that there is a substantial amount of [forestry related] crime that currently does not receive any attention due to manpower and budget restraints.[17] Ministry of Environment, Lands and Parks staff, who play a lesser role in monitoring and enforcement of the code, agree with the statement that "permits are not adequately inspected, monitored, or enforced."[18]

[1] Ministry of Forests, *The British Columbia Forest Practices Code Discussion Paper* (Victoria: Ministry of Forests, 1993). See also: *Forest Practices Code of British Columbia Act*, RSBC 1996, c.159. ss.45(1)(3)(4) and 96(1).

[2] Ministry of Forests-Compliance and Enforcement Branch, *The Annual Report of Compliance and Enforcement Statistics for the Forest Practices Code 1995-1996* (Victoria: Ministry of Forests, 1996), p. 1. Online at: http://www.for.gov.bc.ca/tasb/legsregs/fpc/ann96/Fpc-toc.htm (January 24, 2000).
The Ministry of Environment, Lands and Parks, the Ministry of Energy and Mines, the Ministry of Employment and Investment and the Royal Canadian Mounted Police (primarily through their Forest Crimes Unit) share enforcement obligations with the Ministry of Forests.

[3] Ministry of Forests-Compliance and Enforcement Branch, *The Annual Report of Compliance and Enforcement Statistics for the Forest Practices Code 1995-1996* (Victoria: Ministry of Forests, 1996), pp. 1-2. Online at: http://www.for.gov.bc.ca/tasb/legsregs/fpc/ann96/Fpc-toc.htm (January 24, 2000). See also: Ministry of Forests, *The Annual Report of Compliance and Enforcement Statistics for the Forest Practices Code 1997-1998* (Victoria: Ministry of Forests, 1998). Online at: http://www.for.gov.bc.ca/tasb/legsregs/fpc/ann98/index.htm (January 24, 2000).
There were 31,183 inspections in 1995/1996 and 47,265 inspections reported in 1997/1998.

[4] Ministry of Forests-Compliance and Enforcement Branch, *DRAFT Basic Law (1998) Course Manual* (Victoria: Ministry of Forests, 1998), p. 17.

[5] Keith Moore, "Speaking Notes from a 10-minute Presentation." Paper presented to House of Commons Standing Committee on Natural Resources and Government Operations, Vancouver, BC, May 14, 1999. Online at: http://www.fpb.gov.bc.ca/background/MPs.htm (January 24, 2000).
Keith Moore, Chair of the Forest Practices Board, has also acknowledged that forestry practices have improved since the Code came into effect.

[6] Forest Practices Board, *Logging Plan Approval and Enforcement at Homesite Creek: Complaint Investigation 980142 (FPB/IRC/14)* (Victoria: Forest Practices Board, 1999).

[7] At the time of writing, May 1999, the Forest Practices Board had released 16 audits. This report examined the six audits in which cases of "significant noncompliance" with forestry practices requirements were identified. A case of "significant noncompliance" may constitute a particular event or a significant collection of related events. Cases of "nonsignificant noncompliance" were identified in other audits.

[8] Instances of significant noncompliance in forestry practices were identified by the Forest Practices Board in the following six audits: *South Island Forest District Small Business Forest Enterprise Program -Audit of Forest Planning and Practices; International Forest Products TFL 45 - Audit of Road and Timber Harvesting Practices; Cattermole Timber Forest Licence A19202 - Audit of Operational Planning and Forest Practices; Pretty's Timber Co. Forest Licence A19207 - Timber Harvesting and Road Construction, Maintenance and Deactivation; Plateau Forest Products - Audit of Road and Timber Harvesting Practices; Finlay Forest Industries Inc. Forest Licence A15385 - Audit of Road and Timber Harvesting Practices.*

[9] G.L. Kennah, R.P.F., District Manager Chilliwack Forest District, private communication, June 25, 1999.

[10] Ministry of Forests, *The Annual Report of Compliance and Enforcement Statistics for the Forest Practices Code 1997/1998* (Victoria: Ministry of Forests, 1998), pp. 2-3. Online at: http://www.for.gov.bc.ca/tasb/legsregs/fpc/ann98/comp2.htm (January 24, 2000).

[11] Eric Smith, Corporate Services Director, Vancouver Public Library administration department, private communication, June 2, 1999.

[12] B. Simpson, District Manager, Arrow Forest District, letter to R. Augustin, Kalesnikoff Lumber Co. Ltd., November 30, 1998.

[13] R. Valentine, Area Manager, Ministry of Transportation and Highways letter to C. Pettitt, Regional Coordinator, West Kootenays Forest Watch, June 18, 1999.

[14] Roberta Reader, Director, Ministry of Forests Compliance Enforcement Branch, private communication, July 1, 1999.
Ministry of Forests, Compliance and Enforcement Branch, indicates that utilizing full time equivalents a true assessment of Ministry of Forests expenditures on its compliance and enforcement program work is $32,238,000.
See also: Ministry of Forests, *Annual Report 1996-1997* (Victoria: Ministry of Forests, 1997), Table C-3. Online at: http://www.for.gov.bc.ca/pab/PUBLCTNS/AN_RPTS/9697/table_c3.htm (January 24, 2000).
The Ministry of Forests publicly reports total expenditures on "Monitoring, Enforcement and Audit" of $15,619,000 Ministry of Forests.

[15] Ministry of Forests, *Annual Report 1995/1996* (Victoria: Ministry of Forests, 1996), Table C-3. Online at: http://www.for.gov.bc.ca/pab/publctns/an_rpts/9596an/c-3.htm (January 24, 2000). See also: Ministry of Forests, *Annual Report 1996/1997* (Victoria: Ministry of Forests, 1997), Table C-3. Online at http://www.for.gov.bc.ca/pab/PUBLCTNS/AN_RPTS/9697/table_c3.htm (January 24, 2000).

[16] The only criminal prosecution under the code of a major forest licensee in British Columbia is underway in Penticton against Weyerhaeuser Canada Ltd. There are various enforcement options available to the Ministry of Forests to help it ensure compliance with the law, civil remedies, administrative remedies, quasi-criminal and criminal remedies, criminal prosecutions being the most serious available action.

[17] RCMP Forest Crimes Unit. *Annual Report 1997/1998.*

[18] British Columbia Government, *Environmental protection and management in British Columbia: a report from the men and women who safeguard our environment* (Victoria: British Columbia Government and Service Employees Union, 1999), p. 8.
Ninety percent of respondents to the poll agreed with the statement.

Canada has a National Forest Strategy and Forest Accord

In 1992, Canada's first National Forest Strategy—*Sustainable Forests: A Canadian Commitment*—was released and a National Forest Accord was signed by more than 200 groups.[144] The goal of the strategy was to maintain and enhance the "long-term health of forest ecosystems for the benefit of all living things, both nationally and globally, while providing environmental, economic, social, and cultural opportunities for the benefit of present and future generations."[145] The strategy also includes a commitment to the creation of a set of criteria and indicators to guide sustainable development of forests.

In 1998, a Second National Forest Strategy and Canada Forest Accord (1998-2003) were released. The second National Forest Strategy expands on the promises made in the 1992 version. It commits various governments to conducting broader forest inventories (that is, to include nontimber values), standardizing criteria for testing and monitoring sustainability, and expanding protected areas. In addition, the new strategy requires Canada to pay special attention to the issues of aboriginal (First Nations and Métis) forestry. As part of the Second National Forestry Strategy, participants at the National Forest Congress signed the second Canada Forest Accord. The accord is intended to put into action the vision for sustainable forestry outlined in the National Forest Strategy.[146]

The Blue Ribbon Panel's evaluation of progress in implementing the strategy and moving towards sustainable forest management found that Canada has made some progress in moving towards better forestry practices, but there is still much to be done. *(See below.)*

Canada has a Criteria and Indicators Process

The Canadian Council of Forest Ministers (CCFM) launched the Criteria and Indicators for the conservation and sustainable development of Canada's forests in 1993. The government describes the criteria and indicators as "a toolbox containing concepts and measures to describe sustainable forest management."[147] A framework of domestic criteria and indicators was released in 1995. Thc framework identifies 6 criteria, 22 elements, and 83 indicators.

A 1997 technical report identified a significant number of data gaps that hinder the ability to report on these indicators. In general, the greatest ability to report is in the area of traditional timber harvesting and economic factors. In other areas, national and quantitative data do not exist.[148] The CCFM has agreed to report on Canada's progress toward sustainability to the UN Commission on Sustainable Development in April 2000.[149]

Canada has a Model Forest Network

In June 1992, the government established a network of 10 model forest sites representing 6 of the forest regions. The Canadian Forest Service launched this program to address the challenge of balancing the extensive range of demands on the forest.[150] The initiative focuses on building partnerships locally, nationally, and internationally to generate new ideas and on-the-ground solutions to sustainable forest management issues. A model forest is a partnership between individuals and organizations sharing the common goal of sustainable forest management. There are now 11 model forests. The program also has an international component.[151]

The model forests have been well-received and are considered largely successful. Participants in the National Round Table Rio+5 session noted some process-oriented issues.[152]

Canada has committed 7.6 percent of forests to protected status
8.2 percent of natural regions are adequately protected
40 percent of natural regions are unrepresented

In 1992, Canadian governments committed to completing an ecologically representative network of legally protected areas in Canada by the year 2000 under the Canada Forest Accord.[153] The federal government reports that 83 million hectares, or 9 percent of Canada's total land area, are protected. Of this protected land area,

32 million hectares are forested, representing 7.7 percent of Canada's forested land base.[154]

World Wildlife Fund Canada (WWF) annually reports on nationwide progress toward the year 2000 goal based on a national system of 484 terrestrial natural regions. Of these 484 regions, 388 are forested. WWF reports that 32 of Canada's 388 forested natural regions are adequately represented in protected areas as of July 1999. For these natural regions, protected areas capture the diversity of ecological features characteristic of the region and are of sufficient size and spatial configuration to maintain viable populations of native species and to sustain natural processes.[155]

According to WWF, a further 75 forested regions are moderately represented and 122 regions are partially represented. Forty per cent of Canada's forested natural regions remain unrepresented within protected areas.[156]

Many new protected areas have been established in recent years

Although large gaps in ecological representation remain to be filled in all provinces, some significant progress has been made. For example, in March 1999, the government of Ontario set aside 2.4 million hectares in 378 new protected areas, the largest single expansion of protected areas in Canadian history.[157] One of the largest new parks is Wabakimi Wilderness Park. *(See Box 10.)*

British Columbia protects the Northern Rocky Mountains

Since 1991, the British Columbia government has established more than 200 new parks, including Khutzeymateen Provincial Park, home to the highest known concentrations of grizzly bears on the British Columbia coast; the Northern Rockies; and the Kitlope Heritage Conservancy, the largest intact coastal temperate rainforest in the world.[158] As of April 1998, there were 679 Provincial Parks, Recreation Areas and Ecological Reserves. These total over 10 million hectares, or 10.2 percent of the province.[159]

In 1997, the Province of British Columbia announced that 1.2 million hectares of the Muskwa-Kechika area in the northern Rocky Mountains would be legally protected. Located in northeastern British Columbia, the Muskwa-Kechika contains high concentrations of large mammals, including caribou, elk, moose, Stone's sheep, Plains bison, black bear, and grizzly bear. The British Columbia government's decision to protect the Muskwa-Kechika followed from consensus recommendations that were submitted to government by a local, multistakeholder round table that included members of the public, interest groups, and government, which met over several years to develop a land use plan.[160]

Canada does not yet have endangered species legislation

Canadians also support comprehensive federal legislation to protect endangered species and their habitats. According to polls, 98 percent of Canadians believe the protection of a species habitat is either very (77 percent) or somewhat (21 percent) important. About 9 in 10 Canadians believe the federal government should protect endangered species on all lands in Canada.[161]

The Canadian government has been working with provincial and territorial governments, comanagement boards, aboriginal groups, and various other stakeholders for the last five years to create a national endangered species act. After a failed attempt to develop a bill in the mid-1990s, a new bill is slated to go through the legislative process in the winter of 2000.[162]

VOLUNTARY INITIATIVES

Two certification processes are used in Canada

Certification processes involve the use of independent organizations certifying that products originate from sustainably managed forests. There are two major certification processes in place in Canada: the Canadian Standards Association (CSA) and the Forest Stewardship Council.

BOX 10 Protecting the Caribou of Wabakimi

The need to conserve area-demanding, sensitive species like large ungulates and carnivores is often in conflict with the interests and activities of the forest and mining industries. The province of Ontario, however, showed that it could be done when they expanded Wabakimi Wilderness Park. In the early 1990s, park advocates felt that the 1983 boundaries of Ontario's Wabakimi Wilderness Park were inadequate for the conservation of woodland caribou. This species was supposed to be one of the primary considerations in the park's establishment, since it contained some winter and summer habitat and calving sites.

Most woodland caribou in Ontario live north of the northerly extent of roads and timber harvesting. Wabakimi lies near the southern edge of the line of continuous caribou distribution and has an unusually high concentration of caribou. The size of the caribou population within the immediate region is about 500 animals. Some scientists have indicated that to avoid extinctions in the long term, a minimum of 500 breeding animals may be required. Therefore, despite its large size and relatively high caribou populations, a larger protected area could be at or below the lower limits of long-term caribou viability, despite exchanges of genetic material with populations to the west, east, and north.

From 1993 to 1995, the government of Ontario conducted a decisionmaking process to expand the park. Local and regional representatives were given a mandate to review the existing boundary and develop a single, consensus-based recommendation. The 16-member committee consisted of First Nations, government, tourist outfitter, angler and hunter, conservation organization, mining and forest industry, rural community, and outdoor education representatives.

The committee examined options for a range of values, including preservation of woodland caribou. Other values included traditional use, recreation, economic minerals and timber, tourism, and others. Early on, the committee considered an undefined study area of roughly a million hectares that surrounded the current park of 155,000 hectares. The difficulty of dealing with the complexity of such a vast landscape resulted in defining 60 landscape assessment units of 10,000–50,000 hectares each. Each unit roughly coincided with sub-watersheds. These units were then subjected to a ranking system with respect to "goodness" for different values, including caribou habitat. High-value habitats for lichen-rich winter range, calving areas free from predators, and migration corridors linking winter range and calving areas dominated the assessment of habitat importance for caribou.

The committee generated some options, which were discussed in public forums and reviewed by members of an invited scientific panel. These options incorporated from 45 to 95 percent of the critical caribou habitat considered within the approximately 1.2 million-hectare study area.[1] In 1995, the government of Ontario announced an expansion of Wabakimi Park to an area of approximately 891,500 hectares. This area included 475,000 hectares of critical woodland caribou habitat, or 71 percent of the critical habitat within the larger study area.

[1] P.N Duinker, Ted Armstrong, Bruce Hyer, and Bruce Petersen. 1996. "Using caribou knowledge in expanding the Wabikimi protected area." *Rangifer* Special Issue 10: 183-193.

In early 1994, industry associations from Canada's forest sector requested that the Canadian Standards Association, now CSA-International, create standards for sustainable forest development. CSA-International developed a Sustainable Forest Management (SFM) protocol that requires companies to develop a functioning SFM system that includes public participation. Objectives of the SFM plan must include critical elements of the Canadian Council of Forest Ministers' criteria and indicators process. In addition, companies must adopt an adaptive management style to continually evaluate their impact both on the forest ecosystem and the community. A third-party CSA audit evaluates the SFM system, including an on-site audit of the forest itself. Certification is only valid for five years.[163]

The Forest Stewardship Council (FSC) is an international body that accredits certification organizations. FSC Canada, founded in January 1993, is working on the development of regional performance-based standards to implement and refine global principles. FSC's goal is to promote environmentally responsible, socially beneficial, and economically viable management of the world's forests. To receive certification, a company must adhere to 10 principles of forest stewardship. These principles require companies to obey all applicable laws, respect indigenous peoples, contribute to the well-being of the community, and use the forest in the best economic and environmental fashion. In addition, companies are expected to devise a management plan that will conserve and enhance the ecological health of the forest. The FSC has certified three clients in Canada that own a total of 211,013 hectares of forestland.[164]

A WWF-Canada discussion paper found several problems with the CSA forest certification effort. The key concern relates to the CSA claim that its initiative will result in sustainable forest management. This claim is difficult to make because it involves projections of the impacts of current management practices. WWF also noted some tensions between government representatives and other groups over the extent to which certification might usurp the authority of provincial governments over forest management. WWF noted that the CSA-SFM system did not respect a fundamental principle of standardization, which is that standard-setting activities be separated from auditing against the standard.[165]

Industry has created voluntary programs

In recent years, several industrial associations have created their own voluntary codes of conduct that aim to address, if not surpass, management goals set out in government legislation. One such program of self-assessment is ForestCARE in Alberta. Established by the Alberta Forest Products Association (AFPA), ForestCARE sets standards of performance for all companies that have joined the program. The guiding principles of ForestCARE include sustainable harvests, proper reforestation, watershed protection, and community welfare. About half of AFPA's 67 members now belong to ForestCARE. A 1997 ForestCARE audit of member companies typically reported that "care of the environment" and "care of the forest" exceeded industry standards 90 percent of the time.[166]

50 percent of ForestCARE Companies have been fined for breaking laws

GFW Canada compared the membership of ForestCARE with companies fined by Alberta Environmental Protection for breaking forest-related laws in 1997 and 1998.[167] GFW Canada found that half of all ForestCARE members were fined at least once for breaking the law in those two years. In 1997, 9 of 18 companies fined were members of ForestCARE. *(See Table 15.)*

FINDINGS OF KEY FORESTY REVIEWS

Since 1997, three independent reports have commented on Canada's progress towards sustainable forest management.

The Blue Ribbon Panel

In 1997, the National Forest Strategy Coalition formed a Blue Ribbon Panel of experts to evaluate progress in attaining the 96 commitments in the 1992 National Forest Strategy. The panel's report suggests that Canada has made progress in achieving the goals it outlined in the National Forest Strategy and is moving toward sustainable forest management. But the panel also found that gains

Table 15. Forestry Violations in Alberta

FORESTCARE MEMBERS	VIOLATIONS 1997	FINES 1997	VIOLATIONS 1998	FINES 1998
Alberta-Pacific Forest Industries, Inc.	Excessive road clearing	$17,000	None	
Ainsworth Lumber Co, Ltd.	None		Excessive soil disturbance	$2,000
Blue Ridge Lumber	Oversized topping	$300	None	
Buchanan Lumber	None		Inadequate stream buffer	$7,500
Canadian Forest Products, Ltd.	Unauthorized operations	$500	Damage to watercourse	$3,000
DMI International, Ltd.	None		Removal of buffer on water course	$1,000
High Level Forest Products/DMI	None		Contravention of operating conditions	$800
La Crete Sawmill, Ltd.	Incomplete records	$450	None	
Northland Forest Products, Ltd.	None		Unauthorized timber harvest	$2,000
Spray Lake Sawmills	Failure to reforest	$3,670	None	
Weldwood of Canada, Ltd.	Unauthorized operations	$1,000	None	
Weyerhaeuser Canada, Ltd.	Unauthorized timber/damage to creek	$4,500	Damage to watercourse/damage to watercourse	$6,000
Zeidler Forest Industries, Ltd.	None		Overcutting/contravention of annual plan	$78,510 $1,500

Sources: Alberta Land and Forest Service Contravention Site. Online at: http://www.gov.ab.ca/env/forests/fmd/contra99.html. See also: ForestCARE, *1997 Annual Report* (Edmonton: ForestCARE, 1998). Online at: http://www.abforestprod.org/ar97/8.htm (February 9, 2000).

have been inconsistent across the country and that there is still much work to do. According to the report, 13 percent of the strategy's commitments were fulfilled. There was substantial progress on 38 percent of the commitments, some progress on another 38 percent, little or no progress on 8 percent, and insufficient information to evaluate 2 percent of the commitments. Most gains were made in areas related to timber value rather than nontimber values. Of 13 commitments considered fulfilled, 11 were related to economics, the work force, and global issues. In the two forest stewardship categories, 1 commitment was fulfilled and 9 showed substantial progress. In the forest environment section, 9 of 11 commitments were considered to have made little or some progress.[168]

The panel noted four issues of particular importance that required special effort: completion of an ecological classification of forest lands; completion of a network of protected areas representative of Canada's forest; establishing forest inventories, including nontimber values; and developing a system of national indicators of the sustainability of forest management.[169]

The National Round Table Report

In 1997, the National Round Table on the Environment and the Economy (NRTEE), an independent federal agency with a mandate to promote sustainable development, invited 16 forestry experts (including members from industry, academia, environmental organizations, and aboriginal groups) to discuss progress in Canada's forestry sector since 1992.[170] The group identified both setbacks and progress at the national level and made recommendations for action. In terms of setbacks, participants noted that the federal government had abdicated leadership in forestry policy and had little political will to support sustainable forest management principles. Some participants also felt provinces were handing their responsibilities for sustainable forest management over to industry, which included an abdication of the governments' obligations to First Nations and Métis. This report also found that there was no one standard or even a convergence for sustainable principles across the country and that international agreements were not being implemented at a local level.[171]

In terms of progress, the development of the National Forest Strategy, the creation of Model forests, the Clayoquot Sound Science Panel, Forestry Stewardship Councils, and comanagement agreements are all listed as examples of interesting experiments. The visibility of First Nations has increased in areas such as land use, decisionmaking, and constitutional rights.[172] Participants agreed in general that while progress has been made, there is still much to be done. Recommendations by participants included: developing institutional and regulatory processes to support sustainable forest management; ensuring consistency between international commitments and domestic actions; recognizing and incorporating the value of the traditional ecological knowledge of First Nations and Métis; developing new measures for performance and progress; and including all values when setting harvest levels.[173]

The Senate Subcommittee on the Boreal Forest

In 1999, the federal Senate Subcommittee on the Boreal Forest released a report on Canada's boreal forests. The report states that these forests face real threats and that Canada "must develop strategies that can ensure the survival of our threatened boreal forests while enhancing traditional forest use and preserving economic and industrial benefits."[174] The subcommittee also reports that Canada has "reached the point where potentially transforming concepts are widespread. Yet institutions, management planning, and forestry practices 'on the ground' have not, in most cases, seen significant change."[175] The report recommends developing a system of natural landscape-based forest use regimes that apportion the boreal forest into three distinct management categories. Up to 20 percent of total boreal forest area would be intensively managed for timber and fiber production, while another 20 percent would be set aside as protected areas. The remaining 60 percent would be managed to retain a natural mixture of tree species and ages but also provide some long-term leases that would be audited regularly by community groups assisting forestry experts. This category would attempt to accommodate the full range of forest users and communities.[176]

SECTION 5. CONCLUSION – DATA GAPS AND NEEDS

In this report, we use existing, readily available datasets to document and map the extent of development activities within Canada's forests, the potential implications of these activities in environmental and economic terms, and the key actors (governments, companies, First Nations, and other groups) engaged in forest development.

The maps and graphs presented here offer only partial documentation of trends affecting Canada's forests. However, they provide a baseline for further monitoring and reporting by Global Forest Watch Canada.

We faced four serious obstacles in the preparation of this report:

1. Key datasets are sometimes not collected or not available. As a result, we were unable to adequately quantify a range of values and trends associated with Canada's forests. This included, for example, information on First Nations and Métis values linked to forests; data on logging in primary forests; and complete data (including time series information) on the location of roads.
2. Available data were often outdated, incomplete, and inconsistent.
3. Due to high costs associated with government "cost-recovery" policies, we had limited access to key publicly owned datasets.
4. We had no access to key privately owned datasets. Forest data are often held by companies and are not publicly available.

In this section, we summarize key data challenges faced in compiling this report; identify data gaps (priorities) for better monitoring and reporting on forest development trends in Canada; and outline proposed future Global Forest Watch Canada activities to help fill these data gaps.

DATA LIMITATIONS

Compared to many other countries, Canada has a large amount of information available on its forests. However, much of these data are not readily accessible because governments—as a cost-recovery measure—sell key datasets at prices that are prohibitive for many noncommercial users. As a result, data purchase costs were the greatest constraint we faced in developing maps and indicators for this report. While we could have developed more comprehensive and detailed maps than are presented here, this would have required purchasing provincial ministry resource inventory files, or provincial basemap data, for all provinces. The estimated cost of these datasets ranged up to $6 million.[177] Thus, data costs impose real limitations to environmental reporting.

Government cost-recovery policies are now being questioned from within. In 1997, Natural Resource Canada staff recommended that "the Government of Canada adopt a policy of making federal spatial data available free over the Internet or at the cost of filling a user's request with other media."[178] Their report noted that it is even difficult for government scientists to obtain low-cost, up-to-date spatial data. A group of GIS users are attempting to help change this policy as well, charging that the present government policy "stifles productivity."[179]

Other data limitations GFW Canada faced in preparing this report included:

- Lack of current, accurate, and consistent data (most inventories are at least 10-20 years old, and inventory quality varies greatly between provinces);
- Inability to obtain data in an efficient manner;
- Lack of good metadata (data description) for interpreting and assessing data quality;
- Lack of data standards and lack of consistency among provincial inventories;
- Minimal attribute information (descriptor information on map features); and
- Lack of public access to privately owned inventory data.

KEY DATA GAPS

In general, federal and provincial inventories include limited information on nontimber values associated with forests. Where they exist, these inventories are usually incomplete or too general for useful analysis. This, in turn, limited our ability to assess important questions about the impact of development activities on wildlife, biodiversity, First Nations and Métis and human health. We conducted several analyses to document development trends in Canada. These analyses could have been improved upon with access to additional datasets highlighted below.

- *Forest cover trends:* We documented that while historic and current rates of forest loss are low, some of Canada's most species-rich and productive forests have undergone significant conversion to make way for agricultural and residential areas. How these forests are changing today is unknown. Periodic mapping of forest cover for the entire country would be useful for identifying spatial patterns of forest cover change.

- *Forest development trends:* Data assembled by Global Forest Watch Canada indicate that most of Canada's forests are close to development. Even in its northernmost reaches, Canada's forests are being opened up by roads, mining, and hydroelectric development. This conclusion is based on a nine-month effort to purchase and integrate the best available datasets we could afford on access routes and other development infrastructure. However, in most places we have underestimated the extent of development. For example, we lacked complete information on mine locations and had access to very limited data on hydroelectric development. As a result, we were only able to depict the location of hydroelectric dams rather than the considerable area flooded behind these dams. In most provinces, up-to-date, comprehensive information on road development was either not available or unreasonably costly; as a result, our analysis relied on datasets that underestimated accessible forest area. Road data are used in many countries to identify wilderness areas and as an indicator of fragmentation and potential impacts on biodiversity through access-related human activities. Improved access data (including time series information on road coverage) would be useful for monitoring forest condition relative to a host of nonextractive values derived from forests.

- *Forest industry trends:* As this report shows, logging is the most widespread development activity occurring within Canada's forests. Tenure analysis presented in this report is based on a dataset of forest leases, which was originally developed by World Wildlife Fund Canada. While we have done some updating of this dataset, inclusion of additional attribute (descriptor) information would be particularly valuable for more detailed analysis of the extent and impact of logging activities in Canada. This would include more detailed information showing where companies are actually operating within tenures and the volume of wood extracted relative to annual allowable cut. Surprisingly, in many provinces, governments do not possess comprehensive datasets identifying all companies' areas of operation on the ground. This information, along with forest cover data distinguishing primary from secondary forest and more up-to-date information on land productivity, would be useful for gauging the sustainability of forestry operations. Such data could be used in national analysis addressing whether timber yields can be maintained over the long term, or for assessing the environmental costs associated with logging (for example, monitoring the loss of old-growth habitat).

- *Economic benefits from the forest industry:* Governments and industry collect a vast amount of data on timber production, sales revenues, and employment. This report presents several indicators that highlight the economic importance of forests as a source of revenues from logging. It is more difficult to assess the actual net economic benefits, however, and where these accrue. In particular, scant information was available on subsidies, which would be useful for measuring how much revenue the public actually derives from the logging of public lands. There was also little information available on the degree to which aboriginals participate in the forestry sector. This is a key question, given the land claims issue in Canada.

- *Legislation and performance:* Section four of this report lists the many initiatives, agreements, and policies established to promote the management and protection of forests. We present results of several independent reviews of forest policy, which indicate a mixed record in implementing policies and laws. However, these reviews do not constitute a comprehensive assessment of government and industry performance. Unfortunately, as case study data compiled by Global Forest Watch partners indicate, federal and provincial governments have cut the budgets of ministries responsible for overseeing environmental protection, monitoring development, and enforcing legislation.
One measure of commitment, if not performance, is capacity to implement legislation, in terms of budgets and staffing available to relevant provincial management agencies. Other performance measures include audits (spot-checks) of compliance, violations recorded by enforcement personnel, and fines dispensed. These provide quantifiable indicators that can be tracked, largely by compiling data available through government records.

GLOBAL FOREST WATCH CANADA MONITORING PRIORITIES

To help fill data gaps, and to get this information in the hands of interested parties, Global Forest Watch Canada will undertake a number of future activities:

- We will develop and improve on the datasets we have assembled to date, and wherever possible, make these widely available. As discussed above, we realize that there are limitations to some of our datasets and we are committed to their continual improvement. In particular, we will strive toward more systematic monitoring of industrial activities within Canada's forests, with an emphasis on spatial datasets, so that these data are more useful for analysis and presentation in map form.

- We will initiate a major new Forest Condition Mapping Project. This effort will include identifying remaining tracts of intact forest important for their biodiversity, carbon sequestration, and wilderness values.

- We will expand local monitoring by partner groups and individuals to provide greater focus on regional and local analysis. Increasingly, this task will emphasize on-the-ground field work to collect primary data. For example, we will compile detailed basemap/inventory data and forest development plans. Such data are useful for assessing industry performance with environmental regulations.

- We will expand into additional geographical areas and into working relationships with a broader range of participants. The review process for this report highlighted the need for GFW Canada to expand to areas of Canada where we are currently not active (for example, the Maritimes) and to work with a broader range of stakeholders.

Many government and industry reviewers have indicated their support for the Global Forest Watch approach and a desire to work together to help improve forest stewardship in Canada. We believe that developing these working relationships is an important step in achieving the objectives of Global Forest Watch.

- We will produce and synthesize data on specific topics related to the development, management, and conservation of Canada's forests that are not adequately captured in this report. This might include data assessments of First Nations' issues, the role of private forestlands, and in-depth profiles of specific provinces and regions.

This report is only the first step in monitoring the state of Canada's forests. Results will be available online at our website (www.globalforestwatch.org)and will be improved over time. Comments and feedback are welcome.

NOTES

[1] Environment Canada, *Sustaining Canada's Forests: Forest Biodiversity SOE Bulletin No. 97-1* (Ottawa: Environment Canada, 1997). Online at: http://www3.ec.gc.ca/~ind/English/For_Bio/default.cfm (January 17, 2000).

[2] Dirk Bryant, D. Nielson, and L. Tangley, *The Last Frontier Forests: Ecosystems and Economies on the Edge* (Washington, DC: World Resources Institute, 1997), p. 21.

[3] Dirk Bryant, D. Nielson, and L. Tangley, *The Last Frontier Forests: Ecosystems and Economies on the Edge* (Washington, DC: World Resources Institute, 1997), p. 20.

[4] Environment Canada, *Sustaining Canada's Forests: Forest Biodiversity SOE Bulletin No. 97-1* (Ottawa: Environment Canada, 1997). Online at: http://www3.ec.gc.ca/~ind/English/Home/default1.htm (January 18, 2000).

[5] Dirk Bryant, D. Nielson, and L. Tangley, *The Last Frontier Forests: Ecosystems and Economies on the Edge* (Washington, DC: World Resources Institute, 1997), p. 20.

[6] Corporate Research Associates Inc., *Tracking Survey of Canadian Attitudes Toward Natural Resources Issues, 1997* (Ottawa: Natural Resources Canada, 1997), p. 36. Online at: http://www.nrcan.gc.ca:80/homepage/graphics/survey_e.pdf (January 18, 2000).

[7] 100 hectares (ha) = 1 square kilometer (km) = 247 acres = 0.4 square miles

[8] Canadian Council of Forest Ministers, *Compendium of Canadian Forestry Statistics, 1996* (Ottawa: Natural Resources Canada, 1997), p. 7.

[9] Canadian Forest Service, *The State of Canada's Forests: 1998-99 Innovation* (Ottawa: Natural Resources Canada, 1999), p. 26. Online at: http://nrcan.gc.ca/cfs/proj/ppiab/sof/common/latest.shtml (January 18, 2000).

[10] Dirk Bryant, D. Nielson, and L. Tangley, *The Last Frontier Forests: Ecosystems and Economies on the Edge* (Washington, D.C.: World Resources Institute, 1997), p. 21.

[11] Canadian Council of Forest Ministers, *Compendium of Canadian Forestry Statistics, 1996* (Ottawa: Natural Resources Canada, 1997), pp. 8, 9.

[12] Erin Kellogg, ed. *Coastal Temperate Rain Forests: Ecological Characteristics, Status and Distribution Worldwide* (Portland: Ecotrust and Conservation International, 1992), p. 13.

[13] Erin Kellogg, ed. *Coastal Temperate Rain Forests: Ecological Characteristics, Status and Distribution Worldwide* (Portland: Ecotrust and Conservation International, 1992), p. 3.

[14] Erin Kellogg, ed. *Coastal Temperate Rain Forests: Ecological Characteristics, Status and Distribution Worldwide* (Portland: Ecotrust and Conservation International, 1992), p. 25.

[15] Erin Kellogg, ed. *Coastal Temperate Rain Forests: Ecological Characteristics, Status and Distribution Worldwide* (Portland: Ecotrust and Conservation International, 1992), pp. 3, 7, 9.

[16] Keith Moore, *Coastal Watersheds: An Inventory of Watersheds in the Coastal Temperate Forests of British Columbia* (Vancouver: Earthlife Canada Foundation, 1991), p. 3.
Given the pace of forest development in British Columbia, this report may significantly over-state the status of remaining undeveloped watersheds as of 1999.

[17] BC Ministry of Environment, Lands and Parks. Online at: www.elp.gov.bc.ca/bcparks/explore/parkpgs/kitlope.htm (January 19, 2000).

[18] Natural Resources Defense Council. "Logging Threatens Great Bear Rainforest." Online at: http://www.nrdc.org/status/fogbsr.html (January 19, 2000).

[19] Natural Resources Defense Council. "Logging Threatens Great Bear Rainforest." Online at: http://www.nrdc.org/status/fogbsr.html (January 19, 2000).

[20] Canadian Council of Forest Ministers, *Criteria and Indicators of Sustainable Forest Management in Canada: Technical Report 1997* (Ottawa: Natural Resources Canada, 1997), p. 9. Online at: http://www.nrcan.gc.ca/cfs/proj/ppiab/ci/tech_e.html (January 18, 2000).

[21] United States Department of Agriculture-Forest Service, *Forest Service Roads: A Synthesis of Scientific Information.* (Washington, DC: USDA-Forest Service, 1999), p. 49. Online at: http://www.fs.fed.us/news/roads. (January 14, 2000).

[22] Canadian Council of Forest Ministers, *Criteria and Indicators of Sustainable Forest Management in Canada: Technical Report 1997* (Ottawa: Natural Resources Canada, 1997), p. 65. Online at: http://www.nrcan.gc.ca/cfs/proj/ppiab/ci/tech_e.html (January 18, 2000).

[23] Donald C.E. Robinson, W.A. Kurz, C. Pinkam, *Estimating the Carbon Losses from Deforestation in Canada* (Ottawa: National Climate Change Secretariat, 1999), p. 46.

[24] World Wildlife Fund, *Forest Management Areas and Gaps in Ecological Representation—Analysis Description* (Toronto: WWF-Canada, 1999).

[25] Canadian Council of Forest Ministers, *Criteria and Indicators of Sustainable Forest Management in Canada: Technical Report 1997* (Ottawa: Natural Resources Canada, 1997), p. 1. Online at: http://www.nrcan.gc.ca/cfs/proj/ppiab/ci/tech_e.html (January 18, 2000). See also: Environment Canada, *Sustaining Canada's Forests: Foreset Biodiversity SOE Technical Supplement No. 97-1* (Ottawa: Environment Canada, 1997). Online at: http://www3.ec.gc.ca/~ind/English/Home/default1.htm (January 18, 2000).

[26] Committee on the Status of Endangered Wildlife in Canada (COSEWIC). Online at: http://www.cosewic.gc.ca (January 18, 2000).

[27] Environment Canada, "News Release: Status of six species, including Peregrine Falcon, improves" (Ottawa: Environment Canada, May 7, 1999). Online at: http://www.ec.gc.ca/press/csemdc_n_e.htm (January 18, 2000).

[28] Canadian Forest Service, *The State of Canada's Forests: 1998-99 Innovation* (Ottawa: Natural Resources Canada, 1999), p. 26. Online at: http://nrcan.gc.ca/cfs/proj/ppiab/sof/common/latest.shtml (January 18, 2000).

[29] Canadian Council of Forest Ministers, *Criteria and Indicators of Sustainable Forest Management in Canada: Technical Report 1997* (Ottawa: Natural Resources Canada, 1997), p. 11. Online at: http://www.nrcan.gc.ca/cfs/proj/ppiab/ci/tech_e.html (January 18, 2000).

[30] Canadian Council of Forest Ministers, *Criteria and Indicators of Sustainable Forest Management in Canada: Technical Report 1997* (Ottawa: Natural Resources Canada, 1997), p. 5. Online at: http://www.nrcan.gc.ca/cfs/proj/ppiab/ci/tech_e.html (January 18, 2000).

[31] Conservation Data Centre, Ministry of Environment, Lands and Parks, private communication, November 1999.

[32] Canadian Council of Forest Ministers, *Criteria and Indicators of Sustainable Forest Management in Canada: Technical Report 1997* (Ottawa: Natural Resources Canada, 1997), p. 11. Online at: http://www.nrcan.gc.ca/cfs/proj/ppiab/ci/tech_e.html (January 18, 2000).

[33] Canadian Council of Forest Ministers, *Criteria and Indicators of Sustainable Forest Management in Canada: Technical Report 1997* (Ottawa: Natural Resources Canada, 1997), p. 11. Online at: http://www.nrcan.gc.ca/cfs/proj/ppiab/ci/tech_e.html (January 18, 2000).

[34] Canadian Council of Forest Ministers, *Criteria and Indicators of Sustainable Forest Management in Canada 1997* (Ottawa: Natural Resources Canada, 1997), p. 16. Online at: http://www.nrcan.gc.ca/cfs/proj/ppiab/ci/tech_e.html (January 18, 2000).
These numbers refer to a reduction of at least 50% in the range of a given species compared with its historic range within Canada.

[35] Environment Canada, *Sustaining Canada's Forests: Forest Biodiversity SOE Bulletin No. 97-1* (Ottawa: Environment Canada, 1997), Online at: http://www3.ec.gc.ca/~ind/English/For_Bio/Bulletin/fbind5_e.cfm (January 18, 2000).

[36] Canadian Forest Service, *The State of Canada's Forests: 1998-99 Innovation* (Ottawa: Natural Resources Canada, 1999), p. 26. Online at: http://nrcan.gc.ca/cfs/proj/ppiab/sof/common/latest.shtml (January 20, 2000).

[37] Canadian Forest Service, *The State of Canada's Forests: 1998-99 Innovation* (Ottawa: Natural Resources Canada, 1999), p. 26. Online at: http://nrcan.gc.ca/cfs/proj/ppiab/sof/common/latest.shtml (January 20, 2000).

[38] PriceWaterhouseCoopers, *The Forest Industry in Canada* (Vancouver: PriceWaterhouseCoopers, 1998), p. 13.

[39] Canadian Forest Service, *The State of Canada's Forests: 1998-99 Innovation* (Ottawa: Natural Resources Canada, 1999), p. 26. Online at: http://nrcan.gc.ca/cfs/proj/ppiab/sof/common/latest.shtml (January 18, 2000).

[40] Canadian Forest Service, *The State of Canada's Forests: 1998-99 Innovation* (Ottawa: Natural Resources Canada, 1999), p. 26. Online at: http://nrcan.gc.ca/cfs/proj/ppiab/sof/common/latest.shtml (January 18, 2000).

[41] Environment Canada, *Sustaining Canada's Forests: Timber Harvesting SOE Bulletin No. 99-4* (Ottawa: Environment Canada, 1999), p. 1. Online at: http://www3.ec.gc.ca/~ind/English/Home/default1.htm (January 29, 2000).

[42] Canadian Council of Forest Ministers, "National Forestry Database Program: Revenues and Economic Profile." Online at: http://nfdp.ccfm.org/frames2_e.htm (January 29, 2000).

[43] Canadian Forest Service, *The State of Canada's Forests: 1998-99 Innovation* (Ottawa: Natural Resources Canada, 1999), p. 26. Online at: http://nrcan.gc.ca/cfs/proj/ppiab/sof/common/latest.shtml (January 18, 2000). See also: Statistics Canada, "Exports of Goods on a balance-of-payments basis" and "Imports of Goods on a balance-of –payments basis." Online at: http://www.statcan.ca/english/Pgdb (January 10, 2000).

[44] I.J. Bourke and J. Leitch, *Trade Restrictions and Their Impact on International Trade in Forest Products* (Rome: Food and Agriculture Organization of the United Nations, 1998). (Cited in N. Sizer, D. Downes and D. Kaimowitz. *Tree Trade: Liberalization of International Commerce in Forest Products: Risks and Opportunities* (Washington, DC: World Resources Institute, 1999).).

[45] Canadian Forest Service, *The State of Canada's Forests: 1998-99 Innovation* (Ottawa: Natural Resources Canada, 1999), p. 26. Online at: http://nrcan.gc.ca/cfs/proj/ppiab/sof/common/latest.shtml (January 18, 2000).

[46] Canadian Forest Service, *The State of Canada's Forests: 1998-99 Innovation* (Ottawa: Natural Resources Canada, 1999), p. 26. Online at: http://nrcan.gc.ca/cfs/proj/ppiab/sof/common/latest.shtml (January 18, 2000).

[47] Canadian Forest Service, *The State of Canada's Forests: 1998-99 Innovation* (Ottawa: Natural Resources Canada, 1999), pp. 29, 31. Online at: http://nrcan.gc.ca/cfs/proj/ppiab/sof/common/latest.shtml (January 18, 2000).

[48] Canadian Forest Service, *The State of Canada's Forests: 1998-99 Innovation* (Ottawa: Natural Resources Canada, 1999), p. 26. Online at: http://nrcan.gc.ca/cfs/proj/ppiab/sof/common/latest.shtml (January 18, 2000).
Federal lands in the Northwest Territories (NWT) are managed by the territorial government.

[49] Canadian Forest Service, *The State of Canada's Forests: 1998-99 Innovation* (Ottawa: Natural Resources Canada, 1999), p. 26. Online at: http://nrcan.gc.ca/cfs/proj/ppiab/sof/common/latest.shtml (January 18, 2000). See also: Canadian Forest Service, *The State of Canada's Forests: The People's Forests 1997-98.* (Ottawa: Natural Resources Canada, 1998), pp. 40-41. Online at: http://nrcan.gc.ca/cfs/proj/ppiab/sof/sof98/toc.shtml (January 19, 2000)
Private forest land is either categorized as private woodlots (forested properties owned by individuals, families, municipalities or companies) or industrial freehold (owned by larger corporations for commercial forestry purposes). These private lands are some of the most productive forests in Canada, accounting for 10% of all forest land capable of producing commercial timber and 19% of all industrial roundwood. In terms of other forest products sectors, they account for 77% of maple products, 79% of fuelwood and firewood, and almost 100% of Christmas trees.

[50] Canadian Council of Forest Ministers. "National Forestry Database Program: Terminology." Online at: http://nfdp.ccfm.org/cp95/text_e/sect8t8e.htm (January 19, 2000). See also: K.L. Runyon, *Canada's Timber Supply: Current Status and Outlook, Information Report E-X-45* (Ottawa: Forestry Canada, 1991), pp. 30-31.

[51] In some provinces, such as British Columbia, the AAC is also the minimum rate of cut.

[52] See Elizabeth May, *At the Cutting Edge: The Crisis in Canada's Forests* (Toronto: Key Porter Books, 1998), pp. 33-38 regarding problems with AAC calculations in particular provinces.

[53] Canadian Council of Forest Ministers. "National Forestry Database Program: Annual Allowable Cut: Introduction and Background." Online at: http://nfdp.ccfm.org (January 19, 2000).

[54] Senate Subcommittee on the Boreal Forest, *Competing Realities: The Boreal Forest at Risk* (Ottawa: Senate Standing Committee on Agriculture and Forestry, 1999), pp. 28-29.

[55] Canadian Forest Service, *The State of Canada's Forests: 1998-99 Innovation* (Ottawa: Natural Resources Canada, 1999), p. 4. Online at: http://nrcan.gc.ca/cfs/proj/ppiab/sof/common/latest.shtml (January 18, 2000).

[56] Canadian Council of Forest Ministers, *Compendium of Canadian Forestry Statistics, 1996* (Ottawa: Natural Resources Canada, 1997), p. 118.

[57] Canadian Forest Service, *The State of Canada's Forests: 1998-99 Innovation* (Ottawa: Natural Resources Canada, 1999), p. 26. Online at: http://nrcan.gc.ca/cfs/proj/ppiab/sof/common/latest.shtml (January 18, 2000).

[58] Canadian Council of Forest Ministers. "National Forestry Database Program." Online at: http://nfdp.ccfm.org (January 19, 2000).

[59] Environment Canada, *Sustaining Canada's Forests: Timber Harvesting SOE Bulletin 95-4* (Ottawa: Environment Canada, 1995). Online at: http://www3.ec.gc.ca/~ind/English/Home/default1.htm (January 19, 2000).

[60] For data from 1920-1992, source is: Environment Canada, *Sustaining Canada's Forests: Timber Harvesting SOE Technical Supplement No. 95-4* (Ottawa: Environment Canada, 1995). Online at: http://www3.ec.gc.ca/~ind/English/Home/default1.htm (January 19, 2000).
For data from 1993-1996, source is: Canadian Council of Forest Ministers, "National Forestry Database Program: Table 6.1.A."Online at: http://nfdp.ccfm.org (January 18, 2000).

[61] Canadian Council of Forest Ministers. "National Forestry Database Program: Forest Insects: Quick Facts and Table 4.1." Online at: http://nfdp.ccfm.org (January 19, 2000).

[62] Canadian Council of Forest Ministers, *Criteria and Indicators of Sustainable Forest Management in Canada: Technical Report 1997* (Ottawa: Natural Resources Canada-Canadian Forest Service, 1997), p. 29. Online at: http://www.nrcan.gc.ca/cfs/proj/ppiab/ci/tech_e.html (January 18, 2000).

[63] Canadian Council of Forest Ministers. "National Forestry Database Program: Forest Fires: Tables 3.1a and 3.1b." Online at: http://nfdp.ccfm.org (January 28, 2000).

[64] Canadian Council of Forest Ministers, *Compendium of Canadian Forestry Statistics, 1996* (Ottawa: Canadian Council of Forest Ministers, 1997), p. 42.

[65] Canadian Council of Forest Ministers, *Compendium of Canadian Forestry Statistics, 1996* (Ottawa: Natural Resources Canada, 1997), p. 27.

[66] Tom Clark, *Timber Supply and Endangered Spaces: A World Wildlife Fund Canada Discussion Paper* (Toronto: World Wildlife Fund Canada, 1995), p. 16. See also: Ted Mosquin, P. Whiting and D. McAllister, *Canada's Biodiversity: The Variety of Life, Its Status, Economic Benefits, Conservation Costs and Unmet Needs* (Ottawa: The Canadian Centre for Biodiversity, 1995), p. 71.

[67] B.C. Ministry of Forests – Timber Supply Branch, *AAC listing by Tree Farm Licence and Region and AAC Listing by Timber Supply Area and Region* (Victoria: B.C. Ministry of Forests, 1998). Online at: http://www.for.gov.ba.ca/tsb/other.htm.

[68] M.P. Marchak, S. Aycock, and D.M. Herbert, *Falldown: Forest Policy in British Columbia.* (Vancouver: David Suzuki Foundation and Ecotrust Canada, 1999), pp. 28-29.

[69] K.P. Timoney, *The Old-Growth Forests of Alberta* (Sherwood Park, Alberta: Treeline Ecological Research, 1998), p. 49.

[70] K.L. Runyon, *Canada's Timber Supply: Current Status and Outlook, Information Report E-X-45* (Ottawa: Forestry Canada, 1991), p. 45.

[71] J.P. Kimmins, "Sustainable Development in Canadian Forestry in the Face of Changing Paradigms." Paper presented at Timber Supply in Canada: Challenges and Choices, Kananaskis, Alberta, November 16-18, 1994. (Cited in: Sten Nilsson et al. *How Sustainable are North American Wood Supplies?* (Laxenburg: International Institute for Applied Systems Analysis, 1999), Section 4.6.)

[72] C.L. Wedeles, D.C. Van Damme, and L. Sully. *Alternative silvicultural Systems for Ontario's Boreal Mixed Woods* (Sault Ste. Marie: Canadian Forest Service, 1995). (Cited in: Sten Nilsson et al. *How Sustainable are North American Wood Supplies?* (Laxenburg: International Institute for Applied Systems Analysis, 1999), Section 4.6.)

[73] J. Williams, P. Duinker and G. Bull. *Implications of Sustainable Forest Management for Global Fibre Supply: Working Paper No. 3* (Rome: Food and Agriculture Organization of the United Nations, 1997), p. 43.

[74] D. Williams and J. Tanz, "Summary and analysis of provincial and territorial timber supply status reports." Paper presented at Timber Supply in Canada: Challenges and Choices, Kananaskis, Alberta, November 16-18, 1994. (Cited in Tom Clark, *Timber Supply and Endangered Spaces: A World Wildlife Fund Canada Discussion Paper* (Toronto: World Wildlife Fund Canada, 1995), p. 7.).

[75] Environment Canada, *Sustaining Canada's Forests: Timber Harvesting SOE Bulletin No. 99-4* (Ottawa: Environment Canada, 1999). Online at: http://www3.ec.gc.ca/~ind/English/Home/default1.htm (January 29, 2000).

[76] Ted Mosquin, P. Whiting and D. McAllister, *Canada's Biodiversity: The Variety of Life, Its Status, Economic Benefits, Conservation Costs and Unmet Needs* (Ottawa: The Canadian Centre for Biodiversity, 1995), p. 71.

[77] Environment Canada, *Sustaining Canada's Forests: Forest Biodiversity SOE Bulletin 97-1* (Ottawa: Environment Canada, 1997). Online at: http://www3.ec.gc.ca/~ind/English/Home/default1.htm (January 30, 2000).

[78] Ted Mosquin, P. Whiting and D. McAllister, *Canada's Biodiversity: The Variety of Life, Its Status, Economic Benefits, Conservation Costs and Unmet Needs* (Ottawa: The Canadian Centre for Biodiversity, 1995), p. 72.

[79] Senate Subcommittee on the Boreal Forest, *Competing Realities: The Boreal Forest at Risk* (Ottawa: Standing Senate Committee on Agriculture and Forestry, 1999), pp. 15-17. Online at: http://www.parl.gc.ca/36/1/parlbus/commbus/senate/com-e/bore-e/rep-e/rep09jun99-e.htm (January 19, 2000).The Ontario figures are based on a presentation to the Committee by the Director of the Forest Management Branch, Ontario Ministry of Natural Resources.

[80] Alberta Pacific Forest Industries Inc. Online at: http://www.alpac.ca/ap_story.html (January 19, 2000).

[81] Canadian Council of Forest Ministers, *Criteria and Indicators of Sustainable Forest Management in Canada: Technical Report 1997* (Ottawa: Natural Resources Canada, 1997), p. 38. Online at: http://www.nrcan.gc.ca/cfs/proj/ppiab/ci/tech_e.html (January 18, 2000).

[82] British Columbia Ministry of Forests, *Forest, Range and Recreation Resource Analysis*, (Victoria: British Columbia Ministry of Forests, 1994), p. 72.

[83] Environment Canada, *Sustaining Canada's Forests: Timber Harvesting SOE Bulletin 99-4* (Ottawa: Environment Canada, 1999). Online at: http://www3.ec.gc.ca/~ind/English/Home/default1.htm (January 19, 2000).

[84] Canadian Forest Service, *The State of Canada's Forests: 1998-99 Innovation* (Ottawa: Natural Resources Canada, 1999), p. 26. Online at: http://nrcan.gc.ca/cfs/proj/ppiab/sof/common/latest.shtml (January 18, 2000).

[85] Canadian Council of Forest Ministers, *Criteria and Indicators of Sustainable Forest Management in Canada: Technical Report 1997* (Ottawa: Natural Resources Canada-Canadian Forest Service, 1997), p. 38. Online at: http://www.nrcan.gc.ca/cfs/proj/ppiab/ci/tech_e.html (January 18, 2000).

[86] Canadian Forest Service, *The State of Canada's Forests: 1998-99 Innovation* (Ottawa: Natural Resources Canada, 1999), pp. 23-27. Online at: http://nrcan.gc.ca/cfs/proj/ppiab/sof/common/latest.shtml (January 18, 2000).

[87] Canadian Council of Forest Ministers, *Criteria and Indicators of Sustainable Forest Management in Canada: Technical Report 1997* (Ottawa: Natural Resources Canada-Canadian Forest Service, 1997), pp. 110-111. Online at: http://www.nrcan.gc.ca/cfs/proj/ppiab/ci/tech_e.html (January 18, 2000).

[88] Statistics Canada. "Full time, full-year workers, percentage by earnings class, average and median earnings." Online at: http://www.statcan.ca/english/Pgdb/People/Labour/labor02a.htm (January 18, 2000). See also: PriceWaterhouseCoopers, *The Forest Industry in Canada* (Vancouver: PriceWaterhouseCoopers, 1998), p. 12.

[89] Canadian Forest Service, *The State of Canada's Forests: The People's Forests 1997-98*. (Ottawa: Natural Resources Canada, 1998), p. 80. Online at: http://nrcan.gc.ca/cfs/proj/ppiab/sof/sof98/toc.shtml (January 19, 2000).

[90] Minister of Indian Affairs and Northern Development, *First Nations in Canada* (Ottawa: Minister of Indian Affairs and Northern Development, 1996). Online at: http://www.inac.gc.ca/pubs/fnic/prog.html#peop (January 28, 2000).

[91] Assembly of First Nations-Treaty Relations Unit, *Framework for Action on Treaty Implementation and Enforcement* (Ottawa: Assembly of First Nations, 1999), p. 2.
The Assembly of First Nations, which represents 633 First Nations communities, believes that these treaties uphold the following principles; treaties are based on the sacred and traditional laws of Aboriginals; treaties with Canada were intended to allow for peaceful co-existence with foreign settlers; treaties did not surrender title or ownership of the land and resources; and only nations make treaties.

[92] University of Alberta. "Aboriginal Claims in Canada." Online at: http://www.ualberta.ca/~esimpson/claims/introduction.htm (January 28, 2000).
This website provides an overview of claims issues.

[93] British Columbia Treaty Commission, *1999 Annual Report of the British Columbia Treaty Commission* (Vancouver: British Columbia Treaty Commission, 1999). Online at: http://www.bctreaty.net/publications/99summary.html

[94] Patricia Marchak, S. Aycock and D.M. Herbert, *Falldown: Forestry Policy in British Columbia* (Vancouver: David Suzuki Foundation and Ecotrust Canada, 1999), p. 121.

[95] British Columbia Treaty Commission, *1999 Annual Report of the British Columbia Treaty Commission* (Vancouver: British Columbia Treaty Commission, 1999), p. 2. Online at: http://www.bctreaty.net/publications/99summary.html

[96] Kelly Jane Torrance, "Sovereignty under siege," *Alberta Report* (October 11, 1999): 14-15.

[97] Canadian Forest Service, *The State of Canada's Forests: The People's Forests 1997-98* (Ottawa: Natural Resources Canada, 1998), p. 80. Online at: http://nrcan.gc.ca/cfs/proj/ppiab/sof/sof98/toc.shtml (January 19, 2000).

[98] Jamie Honda-McNeil, Negotiator and Policy Advisor, Alberta Environment, private communication.

[99] P. Marchak, S. Aycock, and D.M. Herbert, *Falldown: Forest Policy in British Columbia* (Vancouver: David Suzuki Foundation and Ecotrust Canada, 1999), p. 83.

[100] Sten Nilsson et al, *How Sustainable Are North American Wood Supplies?* (Laxenburg: International Institute for Applied Systems Analysis, 1999), p. 19.

[101] The Institute on Governance, *Exploring the Relationship Between Aboriginal Peoples and the Canadian Forest Industry: Some Industry Perspectives*, (Ottawa: Institute On Governance, 1998), Section 2.3.3. Online at: http://www.pfpn.gc.ca/sectione/5section/exploring.html#can (January 28, 2000).

[102] Garry Merkel et al, *The Forest Industry's Relationship with Aboriginal Peoples* (Thunder Bay: KBM Consultants, February 1994), pp. 1-3.

[103] The Institute on Governance, *Exploring the Relationship Between Aboriginal Peoples and the Canadian Forest Industry: Some Industry Perspectives*, (Ottawa: Institute On Governance, 1998), Section 2.3.6. Online at: http://www.pfpn.gc.ca/sectione/5section/exploring.html#can (January 19, 2000).

[104] Harry Bombay. 1998. "The Aboriginal Interest Has Been Established." *The Forestry Chronicle* 74(3): 278-282.

[105] The Institute on Governance, *Exploring the Relationship Between Aboriginal Peoples and the Canadian Forest Industry: Some Industry Perspectives*, (Ottawa: Institute On Governance, 1998), Section 2.3.5. Online at: http://www.pfpn.gc.ca/sectione/5section/exploring.html#can (January 28, 2000).

[106] G. Merkel, F. Osendarp and P. Smith, *An analysis of the forest industry's views of Aboriginal participation* (Ottawa: Royal Commission on Aboriginal Peoples, 1994). (Cited in: Doug Brubacher. 1998. "Aboriginal forestry joint ventures: elements of an assessment framework," *The Forestry Chronicle* 74(3): 353-358.)

[107] Commissioner of the Environment and Sustainable Development, *1998 Report of the Commissioner of the Environment and Sustainable Development* (Ottawa: Office of the Auditor General, 1999), Section 2.33. Online at: http://www.oag-bvg.gc.ca/domino/reports.nsf/html/c802ce.html (January 24, 2000).
More information about 48 of Canada's agreements can be found online at: http://www.oag-bvg.gc.ca/domino/env-commitments.nsf/homepage (January 24, 2000).

[108] Canadian Council of Forest Ministers, *Criteria and Indicators of Sustainable Forest Management in Canada: 1997 Progress to Date* (Ottawa: Natural Resources Canada, 1997), p. ii. Online at: http://www.nrcan.gc.ca/cfs/proj/ppiab/ci/prog_e.html (January 20, 2000).

[109] Canadian Forest Service, *The State of Canada's Forests: Learning from History 1996-97* (Ottawa: Natural Resources Canada, 1997), p. 12. Online at: http://nrcan.gc.ca/cfs/proj/ppiab/sof/common/prev.shtml (January 20, 2000). See also Canadian Forest Service, "Canada Supports an International Forest Convention." Online at: http://www.nrcan.gc.ca/cfs/proj/ppiab/for-conv_e.html (January 23, 2000).

[110] International Institute for Sustainable Development. "Briefing: Final Meeting of the Costa Rica-Canada Initiative 6-10 December, 1999: Ottawa Canada." Online at: http://www.iisd.ca/sd/crci/final (January 23, 2000).

[111] Commissioner of the Environment and Sustainable Development. May 1998. "1998 Report Press Conference." Online at: http://www.oag-bvg.gc.ca/domino/other.ndf/html/98coe_e.html (January 26, 2000).

[112] Commissioner of the Environment and Sustainable Development. "Responding to Climate Change – Time to Rethink Canada's Implementation Strategy." Online at: http://www.oag-bvg.bc.ca/domino/reports.nsf/html/c803sume.html (January 24, 2000). See also: "Climate Change Federal government record: Too little action." Online at: http://www.oag-bvg.gc.ca/domino/media.nsf/html/c8pr_e.html (January 24, 2000).

[113] Commissioner of the Environment and Sustainable Development. "Canada's Biodiversity Clock is Ticking." Online at: http://www.oag-bvg.gc.ca/domino/reports.nsf/html/c804ce.html (January 24, 2000).

[114] National Civil Society Consultation on Forestry Issues for Rio+5, "Part Two: Setbacks Since Rio at the National Level," *National Round Table on the Environment and the Economy Report: Rio+5* (Ottawa, NRTEE, 1997). Online at: http://www.ecouncil.ac.cr/rio/national/reports/america/canada/can-for3.htm (January 24, 2000).

[115] These laws are available online at: http://canada.justice.gc.ca/stable/EN/Laws/Chap/index.html (January 24, 2000).

[116] National Civil Society Consultation on Forestry Issues for Rio+5, "Part Two: Setbacks Since Rio at the National Level," *National Round Table on the Environment and the Economy Report: Rio+5* (Ottawa, NRTEE, 1997). Online at: http://www.ecouncil.ac.cr/rio/national/reports/america/canada/can-for3.htm (January 24, 2000).

[117] Blue Ribbon Panel, *National Forest Strategy "Sustainable Forests: A Canadian Commitment" Final Evaluation Report* (Ottawa: National Forest Strategy Coalition, 1997) Section 1-10.

[118] Canadian Institute of Forestry. 1996. "The Need for a Canadian Federal Presence in Forestry." Online at: http://www.cif-ifc.org/pp2.html (January 26, 2000).

[119] House Standing Committee on the Environment and Sustainable Development, *Report Number Three: Enforcing Canada's Pollution Laws: The Public Must Come First* (Ottawa: House of Commons, May 1998), Paragraph 31. Online at: http://www.parl.gc.ca/InfocomDoc/36/1/ENSU/Studies/Reports/ensurp03-e.htm (January 24, 2000).

[120] House Standing Committee on the Environment and Sustainable Development, *Report Number Three: Enforcing Canada's Pollution Laws: The Public Must Come First* (Ottawa: House of Commons, May 1998), Paragraphs 72-74. Online at: http://www.parl.gc.ca/InfocomDoc/36/1/ENSU/Studies/Reports/ensurp03-e.htm (January 24, 2000).

[121] House Standing Committee on the Environment and Sustainable Development, *Report Number Three: Enforcing Canada's Pollution Laws: The Public Must Come First* (Ottawa: House of Commons, May 1998), Paragraph 110. Online at: http://www.parl.gc.ca/InfocomDoc/36/1/ENSU/Studies/Reports/ensurp03-e.htm (January 24, 2000).

[122] House Standing Committee on the Environment and Sustainable Development, *Report Number Three: Enforcing Canada's Pollution Laws: The Public Must Come First* (Ottawa: House of Commons, May 1998), Paragraph 41. Online at: http://www.parl.gc.ca/InfocomDoc/36/1/ENSU/Studies/Reports/ensurp03-e.htm (January 24, 2000).

[123] House Standing Committee on the Environment and Sustainable Development, *Report Number Three: Enforcing Canada's Pollution Laws: The Public Must Come First* (Ottawa: House of Commons, May 1998), Paragraphs 72-79. Online at: http://www.parl.gc.ca/InfocomDoc/36/1/ENSU/Studies/Reports/ensurp03-e.htm (January 24, 2000).

[124] House Standing Committee on the Environment and Sustainable Development, *Report Number Three: Enforcing Canada's Pollution Laws: The Public Must Come First* (Ottawa: House of Commons, May 1998), Paragraph 113. Online at: http://www.parl.gc.ca/InfocomDoc/36/1/ENSU/Studies/Reports/ensurp03-e.htm (January 24, 2000).

[125] Peter Krahn. 1999. "The Business of Environmental Law." *Global Biodiversity* 8: 2-8. See also Environment Canada. Online at: http://www.pyr.ec.gc.ca/ep/enforcement/envsvo98.htm (January 2000).

[126] Mark Winfield et al, *Ontario's Environment and the Common Sense Revolution: A FourYear Report*, (Toronto: Canadian Institute For Environmental Law and Policy, 1999), Section 4-2.

[127] Mark Winfield et al, *Ontario's Environment and the Common Sense Revolution: A FourYear Report*, (Toronto: Canadian Institute For Environmental Law and Policy, 1999), Section 4-2.

[128] Mark Winfield et al, *Ontario's Environment and the Common Sense Revolution: A FourYear Report*, (Toronto: Canadian Institute For Environmental Law and Policy, 1999), Section 4-3.

[129] Sierra Legal Defence Fund and Wildlands League, "Cutting Around the Rules," (Vancouver: Sierra Legal Defence Fund, April 1998), p. 16.

[130] Environmental Commissioner of Ontario, *1998 Annual Report: Open Doors: Ontario's Environmental Bill of Rights* (Toronto: Environmental Commissioner of Ontario, 1998), p. 186. Online at: http://www.eco.on.ca/english/publicat/ar98/report98.pdf (January 24, 2000).

[131] Mark Winfield et al, *Ontario's Environment and the Common Sense Revolution: A FourYear Report*, (Toronto: Canadian Institute For Environmental Law and Policy, 1999), Section 4-5.

[132] Environmental Commissioner of Ontario, *1998 Annual Report: Open Doors: Ontario's Environmental Bill of Rights* (Toronto: Environmental Commissioner of Ontario, 1998), p. 187. Online at: http://www.eco.on.ca/english/publicat/ar98/report98.pdf (January 24, 2000).

[133] Sierra Legal Defence Fund, *British Columbia Clearcut Code*, (Vancouver: Sierra Legal Defence Fund, November 1996), p. 1. Online at: http://www.sierralegal.org/reports.html (January 24, 2000).

[134] Sierra Legal Defence Fund, *Stream Protection Under The Code*, (Vancouver: Sierra Legal Defence Fund, February 1997), p. ii. Online at: http://www.sierralegal.org/reports.html (January 24, 2000).

[135] Sierra Legal Defence Fund, *British Columbia Forestry Report Card 1997-98*, (Vancouver: Sierra Legal Defence Fund, 1998), pp. 7-8. Online at: http://www.sierralegal.org/reports.html (January 24, 2000).

[136] The three on-the-ground mechanisms intended to protect biodiversity and threatened and endangered wildlife are: Landscape Units, Old-Growth Management Areas and Wildlife Habitat Areas: Sierra Legal Defence Fund, *Wildlife At Risk: The Lack of protection for Endangered Species, Wildlife, and Biodiversity Under the BC Forest Practice Code* (Vancouver: Sierra Legal Defence Fund, 1997), pp. 3-4. Online at: http://www.sierralegal.org/reports.html (January 24, 2000). As of January 30, 2000, fifty-four months since the Code was enacted in law, not one of these on the ground protections have been fully implemented anywhere in BC.

[137] British Columbia Ministry of Environment Lands and Parks, *Greenpeace Report: Broken Promises Analysis*, (Vancouver: Ministry of Environment Lands and Parks, 1997), p. 2.

[138] Sierra Legal Defence Fund, *Going Downhill Fast: Landslides and the Forest Practices Code*, (Vancouver: Sierra Legal Defence Fund, 1997), pp. 3-4. Online at: http://www.sierralegal.org/reports.html (January 24, 2000).

[139] Greenpeace, *Broken Promises: The Truth About What's Happening to British Columbia's Forests*, (Vancouver: Greenpeace, April 1997), p. 16.

[140] Ministry of Forests and Ministry of Environment, Lands and Parks, *Cutblock Riparian Management Area Review*, (Victoria, BC: Ministry of Forests – Vancouver Forest Region and Ministry of Environment, Lands and Parks – Vancouver Island Region, March 21, 1997), p. 1.

[141] British Columbia Forest Practices Board, *Forest Planning and Practices in Coastal Areas and Streams* (Victoria: Forest Practices Board, June 1998). Online at: http://www.fpb.gov.bc.ca/reports/special_invest/techrep1/mainfind.htm#opplan)

[142] British Columbia Forest Practices Board, *Forest Planning and Practices in Coastal Areas and Streams,* (Victoria: Forest Practices Board, June 1998), p. 17. Online at: http://www.fpb.gov.bc.ca/reports/special_invest/techrep1/forplan1.htm

[143] British Columbia Forest Practices Board, *Forest Planning and Practices in Coastal Areas and Streams,* (Victoria: Forest Practices Board, June 1998), Exhibit III-17. http://www.fpb.gov.bc.ca/reports/special_invest/techrep1/mainfind.htm#ex11117

[144] Canadian Council of Forest Ministers, *National Forestry Strategy (1998-2003): Sustainable Forests: A Canadian Commitment*, (Ottawa: Natural Resources Canada, 1998), pp. vii-viii. Online at: http://www.nrcan.gc.ca/cfs/nfs/strateg/final_e.html (January 24, 2000).

[145] Blue Ribbon Panel, *National Forest Strategy "Sustainable Forests: A Canadian Commitment" Final Evaluation Report* (Ottawa: National Forest Strategy Coalition, 1997), p. 5.

[146] Canadian Council of Forest Ministers, *National Forestry Strategy (1998-2003): Sustainable Forests: A Canadian Commitment* (Ottawa: Natural Resources Canada, 1998), p. 5. Online at: http://www.nrcan.gc.ca/cfs/nfs/strateg/final_e.html (January 24, 2000).

[147] Canadian Forest Service, *The State of Canada's Forests: Learning from History 1996-97* (Ottawa: Natural Resources Canada, 1997), p. 68. Online at: http://nrcan.gc.ca/cfs/proj/ppiab/sof/common/prev.shtml (January 20, 2000).

[148] Canadian Council of Forest Ministers, *Criteria and Indicators of Sustainable Forest Management in Canada: Technical Report 1997* (Ottawa, Natural Resources Canada, 1997), p. iii. Online at: http://www.nrcan.gc.ca/cfs/proj/ppiab/ci/tech_e.html (January 18, 2000).

[149] Canadian Forest Service, *The State of Canada's Forests: 1998-99 Innovation* (Ottawa: Natural Resources Canada, 1999), p. 39. Online at: http://nrcan.gc.ca/cfs/proj/ppiab/sof/common/latest.shtml (January 18, 2000).

[150] The Canadian Model Forest Network. Online at: http://www.modelforest.net (January 24, 2000). A history of the development of this program and initial reviews are available at this site.

[151] The International Model Forest Secretariat. Online at: http://www.idrc.ca/imfn (January 24, 2000).

[152] National Civil Society Consultation on Forestry Issues for Rio+5, "Part Three: Progress since Rio at the National Level," *National Round Table on the Environment and the Economy Report: Rio+5* (Ottawa: National Round Table on the Environment and the Economy, 1997). Online at: http://www.ecouncil.ac/rio/national/REPORTS/AMERICA/canada/Can-for4.htm (January 24, 2000).

[153] Canadian Council of Forest Ministers, *Canada Forest Accord* (Ottawa: Natural Resources Canada, 1992). Online at: http://www.nrcan.gc.ca/cfs/nfs/strateg/acontrol_e.html (January 24, 2000).

[154] Canadian Forest Service, *The State of Canada's Forests: 1998-99 Innovation* (Ottawa: Natural Resources Canada, 1999), p. 26, 82. Online at: http://nrcan.gc.ca/cfs/proj/ppiab/sof/common/latest.shtml (January 18, 2000).

[155] World Wildlife Canada, *Canada's Commitment to Forest Protected Areas: A WWF Status Report* (Toronto: WWF-Canada, 1999).

[156] World Wildlife Canada, *Canada's Commitment to Forest Protected Areas: A WWF Status Report* (Toronto: WWF-Canada, 1999).

[157] World Wildlife Canada, *Canada's Commitment to Forest Protected Areas: A WWF Status Report* (Toronto: WWF-Canada, 1999).

[158] British Columbia Ministry of Environment, Lands and Parks. "British Columbia Parks." Online at: http://www.elp.gov.bc.ca/bcparks/facts/facts.htm (January 24, 2000).

[159] British Columbia Land Use Coordination Office. "Protected Areas Strategy Update 1998 (Brochure)" Online at: http://www.luco.bc.gc.ca/pas/pasuptoc.htm (January 26, 2000).

[160] British Columbia Land Use Coordination Office. "Northern Rockies." Online at: http://www.luco.gov.bc.ca/nrockies/home.htm (January 24, 2000).

[161] Pollara Inc. "Canadian Attitudes and Opinions Towards the Protection of Endangered Species." Poll conducted 1999.

[162] The new "Species at Risk Act" proposal is online at: http://www.ec.gc.ca/sara/index.html (January 26, 2000).

[163] Katie Althoff, CSA-International, private communication, December 1999.

[164] Forest Stewardship Council. Online at: http://www.fscoax.org/html/noframes/5-3-3.html (January 24, 2000) .

[165] Chris Elliott and Arlin Hackman, *Current Issues in Forest Certification in Canada* (Toronto: WWF Canada, 1996), pp. 11, 26, 34-35.

[166] Alberta Forest Products Association. *ForestCARE Annual Report 1997* (Edmonton: Alberta Forest Products Association, 1998), p. 3-10. Online at: http://www.abforestprod.org/ar97 (January 24, 2000).

[167] Alberta Land and Forest Service Contravention Site. Online at: http://www.gov.ab.ca/env/forests/fmd/contra/contra98.html#months (January 24, 2000).

[168] Blue Ribbon Panel, *National Forest Strategy "Sustainable Forests: A Canadian Commitment" Final Evaluation Report* (Ottawa: National Forest Strategy Coalition, 1997), Sections 1, 2-26.

[169] Blue Ribbon Panel, *National Forest Strategy "Sustainable Forests: A Canadian Commitment" Final Evaluation Report* (Ottawa: National Forest Strategy Coalition, 1997), p. 1.

[170] National Civil Society Consultation on Forestry Issues for Rio+5, "Background," *National Round Table on the Environment and the Economy Report: Rio+*5 (Ottawa: National Round Table on the Environment and the Economy, 1997), p. 1. Online at: http://www.ecouncil.ac/rio/national/REPORTS/AMERICA/canada/Can-for.htm (January 24, 2000).

[171] National Civil Society Consultation on Forestry Issues for Rio+5, "Part Two: Setbacks Since Rio at the National Level," *National Round Table on the Environment and the Economy Report: Rio+*5 (Ottawa: National Round Table on the Environment and the Economy, 1997), p. 1. Online at: http://www.ecouncil.ac/rio/national/REPORTS/AMERICA/canada/Can-for3.htm (January 24, 2000).

[172] National Civil Society Consultation on Forestry Issues for Rio+5, "Part Three: Progress since Rio at the National Level," *National Round Table on the Environment and the Economy Report: Rio+*5 (Ottawa: National Round Table on the Environment and the Economy, 1997), pp. 1-5. Online at: http://www.ecouncil.ac/rio/national/REPORTS/AMERICA/canada/Can-for4.htm (January 24, 2000).

[173] National Civil Society Consultation on Forestry Issues for Rio+5, "Part Four: Recommendations for Action at the National Level," *National Round Table on the Environment and the Economy Report: Rio+*5 (Ottawa: National Round Table on the Environment and the Economy, 1997) pp. 1-4. Online at: http://www.ecouncil.ac/rio/national/REPORTS/AMERICA/canada/Can-for5.htm (January 24, 2000).

[174] Senate Subcommittee on the Boreal Forest, *Competing Realities: The Boreal Forest at Risk* (Ottawa: Standing Senate Committee on Agriculture and Forestry, 1999) Preface. Online at: http://www.parl.gc.ca/36/1/parlbus/commbus/senate/com-e/rep-e.htm (January 24, 2000).

[175] Senate Subcommittee on the Boreal Forest, *Competing Realities: The Boreal Forest at Risk* (Ottawa: Standing Senate Committee on Agriculture and Forestry, 1999) p. 5. Online at: http://www.parl.gc.ca/36/1/parlbus/commbus/senate/com-e/rep-e.htm (January 24, 2000).

[176] Senate Subcommittee on the Boreal Forest, *Competing Realities: The Boreal Forest at Risk* (Ottawa: Standing Senate Committee on Agriculture and Forestry, 1999) p. ii. Online at: http://www.parl.gc.ca/36/1/parlbus/commbus/senate/com-e/rep-e.htm (January 24, 2000).

[177] This number is based on the following amounts: Newfoundland, $31,550; Nova Scotia, $79,600; Quebec, $150,000 (for MOF paper maps); and British Columbia, $5,972,950 (for both base map and forestry inventory data - need basemap info to register inventory). Note it is possible to purchase the federal Canadian Road Network for approximately $50,000; however it is very old data (1950-1996) that is barely updated anymore.

[178] Goodenough, D.G., D. McKenney and B. Pendrel. 1997. Report on Spatial Data in the United States Government and Recommendations for Natural Resources Canada. Pacific Forest Centre, Natural Resources Canada. Available On line at: http://www.pfc.cfs.nrcan.gc.ca.

[179] Levinsohn, Allan. 1999. "Canadian Geospatial Data Policy Stifles Productivity." Available on line at http://members.home.net/freedata/datachrg.htm (January 26, 2000) or www.geoplace.com/gw/1999/0699/699can.asp (January 26, 2000).

APPENDIX 1. DATASETS AND TECHNICAL NOTES

SOURCE DATA

Maps

FOREST REGIONS OF CANADA

DATA: Forest Regions of Canada (polygon) vector data
PROVIDER: Pacific Forestry Center, Canadian Forest Service, NRCAN, http://www.pfc.cfs.nrcan.gc.ca/
SOURCE DATE: 1972.
SCALE: 1:7,500,000

1995 LAND COVER OF CANADA

DATA: Land Cover of Canada Version 1.1 raster data
PROVIDER: Canadian Center for Remote Sensing (CCRS), EMS Section, http://www.ccrs.nrcan.gc.ca/ccrs
SOURCE DATE: 1995
SCALE: 1 km raster grid

APPROXIMATE COMMERCIAL FOREST ZONE OF CANADA/FORESTRY COMPANIES

NAME: WWF Tenure Database
PROVIDER: Original data obtained from provincial governments by World Wildlife Fund. Original data compiled by World Wildlife Fund, http://www.wwfcanada.org/
SOURCE DATE: Average Fall 1998.
SCALE: Variable. Estimated average scale 1:1,000,000

CANADA LAND INVENTORY LAND CAPABILITY FOR FORESTRY/ONTARIO TIMBER INVENTORY

NAME: Canada Land Inventory Land Capability for Forestry /Ontario Timber Inventory
PROVIDER: Geogratis, Geomatics Canada, Canadian Center for Remote Sensing, NRCAN, http://geogratis.cgdi.gc.ca/frames.html
SOURCE DATE: Inventory was maintained between early 1960 and 1984.
SCALE: 1:250,000 (except BC); BC 1:1,000,000

ACCESS DATA

A variety of data sources were used:

NAME: VMAP0_R4 (Digital Chart of the World) transportation and utility vector data and populated places point data.
PROVIDER: Geogratis, Geomatics Canada Canadian Center for Remote Sensing, NRCAN http://geogratis.cgdi.gc.ca/frames.html
SOURCE DATE: Approximate Average 1990.
SCALE: 1:1,000,000.

DATA: 1:50,000 roads, railways and utilities database.
PROVIDER: Nova Scotia Geomatics Center, http://www.nsgc.gov.ns.ca/
SOURCE DATE: Estimated Average mid-80s to early 90s. Highways more recently updated than secondary roads.
SCALE: 1:50,000

DATA: Digital Topographic Database of New Brunswick.
PROVIDER: Service New Brunswick, http://www.gov.nb.ca/snb/
SOURCE DATE: 1996.
SCALE: 1:10,000

DATA: "Le Quebec forestier meridional" (LANDSAT satellite image composite of Quebec commercial forest zone).
PROVIDER: La Direction de la gestion des stocks forestiers, Ministere des Ressources naturelles du Quebec, http://www.mrn.gouv.qc.ca/intro.asp
SOURCE DATE: 1993/1994
SCALE: 125 meter resolution

DATA: Raster summary grid file of access corridors. Original source data includes:

- Ontario Base Map 1:20,000 roads, railway, and transmission line data;
- Ontario logging roads from Landsat imagery;
- Ontario roads from the Digital Topographic Database 1:600,000 scale;

Ontario Provincial Snowmobile Trails.
PROVIDER: Ontario Ministry of Natural Resources.
SOURCE DATE: Variable – estimated between mid-80s to mid-90s.
SCALE: 200m resolution raster grid
PROCESSING: Projected to *Land Cover of Canada*. 200m data buffered to 1km access corridor. Resampled to 1km resolution. Reclassed to single access class.

DATA: Manitoba Road Network (from Forest Resources Inventory data) roads data.
PROVIDER: Manitoba Natural Resources, Forestry Branch
SOURCE DATE: Quoted average age "10-20 years old".
SCALE: 1:20,000

DATA: Saskatchewan Road Network road and railway data (forested region only)
PROVIDER: SaskGeomatics, Saskatchewan Property Management Corporation, http://www.gov.sk.ca/spmc/sgd/saskgeo.htm
SOURCE DATE: 1996-1998.
SCALE: 1:20,000

DATA: Recreation Opportunity Spectrum
PROVIDER: Original data from BC Ministry of Forests, Recreation Branch, ARC/INFO data obtained from Earthlife Canada Foundation, Vancouver, BC
CREATION DATE: 1989
SCALE: 1:2,000,000

DATA: 1:250,000 topographic data (includes roads, trails, seismic lines and utility lines).
PROVIDER: Yukon Government, Department of Renewable Resources
SOURCE DATE: Variable. Original data 1970-1990. Partially updated 1995.
SCALE: 1:250,000

DATA: 1:20,000 Alberta Provincial Base Maps

DATA: 1:50,000 Alberta Resource Access Maps

DATA: Plot for Alberta section of the Western Canadian Sedimentary Basin
PROVIDER: National Energy Board of Canada
SCALE: 1:100,000

DATA: Aerial photographs
SCALE: 1:40,000

DATA: 1998 Landsat 5 images for portions of Alberta

DATA: Structure and Architecture of the Western Canadian Sedimentary Basin Map
PROVIDER: Alberta Energy and the Geological Survey of Canada
SOURCE DATE: 1994

MINING DATA

NAME: Principal Mineral Areas of Canada (MAP 900A)
PROVIDER: Minerals and Metals Sector, Geological Survey of Canada, NRCAN, http://www.pfc.cfs.nrcan.gc.ca/
SOURCE DATE: 1997
SCALE: 1:6,000,000

NAME: MINSYS (used for past producers in Alberta, Manitoba, and Quebec)
PROVIDER: Minerals and Metals Sector, Geological Survey of Canada, NRCAN, http://www.pfc.cfs.nrcan.gc.ca/
SOURCE DATE: Last compilation estimated 1990.
SCALE: point data, accuracy unknown.

NAME: Mining / Quarry Activity on Public Lands
PROVIDER: Alberta Environment, Land and Forest Service, Land Administration Division
SOURCE DATE: 1999
SCALE: point data, general county location only.

NAME: BC MINFILE
PROVIDER: MINFILE Unit, BC Ministry of Energy and Mines, http://www.em.gov.bc.ca/Mining/Geolsurv/Minfile/default.htm
SOURCE DATE: June 1999
SCALE: Approx. 1:50,000

NAME: Mineral Occurrence Database System (MODS)
PROVIDER: Geological Survey of Newfoundland and Labrador, http://www.geosurv.gov.nf.ca/
SOURCE DATE: Not Available.
SCALE: Not Available.

NAME: New Brunswick Mineral Occurrence Database.
PROVIDER: Minerals and Energy Division, Department of Natural Resources and Energy http://www.gov.nb.ca/dnre/minerals/index.htm
SOURCE DATE: 1992.
SCALE: Point data in DMS, unknown accuracy.

NAME: NORMIN.DB database
PROVIDER: Northwest Territories Region, Mineral Resources Directorate, Geology Division, Department of Indian and Northern Affairs, http://www.inac.gc.ca/regions/nt/geolog.html
SOURCE DATE: 1998
SCALE: Point data in DMS, unknown accuracy.

NAME: Mineral Occurrence Database
PROVIDER: Mineral Inventory Program, Minerals and Energy Branch, Department of Natural Resources, http://www.gov.ns.ca/natr/meb/index.htm
SOURCE DATE: 1999
SCALE: Point data in UTM, unknown accuracy.

NAME: Abandoned Mines database
PROVIDER: Mineral Inventory Program, Minerals and Energy Branch, Department of Natural Resources, http://www.gov.ns.ca/natr/meb/index.htm
SOURCE DATE: 1997
SCALE: Point data in UTM, unknown accuracy.

NAME: Mineral Deposit Inventory (MDI2)
PROVIDER: Ontario Geological Survey, Ministry of Natural Resources, http://www.gov.on.ca/MNDM/MINES/PUB/digcat/min.htm
SOURCE DATE: 1998
SCALE: Approx. 1:50,000

NAME: Mine List
PROVIDER: Saskatchewan Energy and Mines, http://www.gov.sk.ca/cgi-bin/smdi
SOURCE DATE: 1999.
SCALE: point data (UTM), unknown accuracy

NAME: Yukon MINFILE
PROVIDER: Yukon Geology Program, Department of Renewable Resources, http://www.yukonweb.yk.ca/government/geoscience/publications/minfile
SOURCE DATE: 1997
SCALE: point data, unknown accuracy.

HYDROELECTRIC DATA

NAME: "Electric Power Generating Stations", Catalogue 57-206-XPB
PROVIDER: Statistics Canada, http://www.statcan.ca/start.html
SOURCE DATE: 1997
SCALE: point data, general location only.

NAME: Small Hydroelectric Generating Stations
PROVIDER: Energy Sector, Natural Resources Canada, Tabular data available from GeoGratis. Contact Tony Tung.
SOURCE DATE: 1998
SCALE: point data (lat/long).

NAME: 1:2,000,000 Dams
PROVIDER: National Atlas of Canada 1:2,000,000 series. Digital data available from GeoGratis, http://geogratis.cgdi.gc.ca/frames.html
SOURCE DATE: Mid 1980s.

WATERSHED DATA

NAME: Canada Watersheds (Sub-Sub Drainage Basins)
PROVIDER: Federal Geogratis website, http://geogratis.cgdi.gc.ca/frames.html
SCALE: 1:2,000,000
SOURCE DATE: Page: 4
Original source 1972 to 1977, revised mid-80s (from Geogratis doc). This coverage was digitized from a published Environment Canada Map depicting active and discontinued hydrometric stations operated by Water Resources Branch from 1972 to 1977. The coverage depicts primary, secondary and tertiary drainage basins across Canada.

FIRST NATIONS HISTORIC TREATIES

NAME: Historical Treaties of Canada. National Atlas of Canada MCR 4162
PROVIDER: Geomatics Canada, Canadian Center for Remote Sensing, Natural Resources Canada, http://geogratis.cgdi.gc.ca/frames.html
COMPILATION DATE: 1986.
SCALE: 1:7,500,000. All Treaty Boundaries are approximate only. See original map for full description.
PROCESSING: Projected to Lambers Conformal Conic, NAD 27.

FIRST NATIONS SETTLEMENTS AND LAND CLAIMS

NAME: N/A
PROVIDER: Graphic and text-based descriptions of recent Land ClaimAboriginal Settlement Agreements and outstanding land claims. Sources include:

- Comprehensive Land Claims in Canada (11x17 Map). INAC, 1993.
- Comprehensive Land Claims Policy and Status of Claims. INAC, 1999.
- Yukon First Nation Traditional Territories (GIS data). Yukon Department of Renewable Resources, 1998.
- Nunavut Territory. Geogratis, Canadian Center for Remote Sensing, 1998.
- The Natives in Quebec (map). Ministere des Ressources Naturelles, Gouvernement du Quebec, 1996.
- Western Arctic Boundaries. Ministry of Aboriginal Affairs, Government of the Northwest Territories, 1998.
- Labrador Innu and Inuit Land Claims. Ministry of Aboriginal Affairs, Government of Newfoundland and Labrador, 1999.
- 1999 Annual Report. BC Treaty Commission, 1999.

SOURCE DATE: Various.
SCALE: Suitable for graphic display only.
PROCESSING:

- Settlement lands and/or land claim boundaries digitized on-screen using VMAP_R4 political boundaries for outline of settlements/land claims where applicable. All non-political boundaries approximated visually from maps or graphics. Attributes added as required.

FIRST NATION, METIS AND INUIT COMMUNITIES

NAME: Native Reserves
PROVIDER: Original Source Department of Indian and Northern Affairs.
Provided by Legal Surveys Branch, Geomatics Canada
Canadian Center for Remote Sensing, NRCAN
SOURCE DATE: 1998.
SCALE: N/A. Point data only.

PROCESSING:

- Original INAC latitude/longitude point data converted to decimal degrees. In many cases, INAC data was missing or erroneous. In these cases, where explicit town names were provided within the INAC database, latitude/longitude coordinates were obtained by searching the Canadian Geographical Names Data Base (CGNDB) - http://geonames.nrcan.gc.ca/english/.
- In cases where location data was not available from the CGNDB, missing reserves and settlements were located using the National Atlas of Canada "*Canada – Indian and Inuit Communities*" map series (MCR 4025,4026,4028, 4029,4030).

Tables and Figures

NAME: National Forestry Database Program
PROVIDER: Canadian Council of Forest Ministers. Available online at: http://nfdp.ccfm.org
SOURCE DATE: Variable

NAME: Statistical Data
PROVIDER: Statistics Canada. Available online at: http://www.statcan.ca/
SOURCE DATE: Variable

METHODOLOGY AND PREPARATION OF MAPS

MAP 1. CANADA'S FOREST HERITAGE

DEFINITIONS
Forest Regions: defined by Rowe (1972)

MAP 2. DEVELOPMENT STATUS OF FORESTED WATERSHEDS

DEFINITIONS
WATERSHED STATUS: Classes are:

- No known developments
- 0 to 10 percent developed
- 10 to 25 percent developed
- 25 to 50 percent developed
- Over 50 percent developed

INDUSTRIAL DEVELOPMENT: Known locations for:

- Dams: Sum of large generating stations > 20 megawatts (Statistics Canada), small generating stations/dams less than 20 megawatts (small hydro data) and dams from federal 1:2,000,000 basedata.

- Mineral Development:
  - Principal Mining Areas (1997): Operating metal, industrial mineral, and coal mines in 1997.
  - Abandoned Mine: Mineral occurrences with Past Producer status.

- Settlements: Built-up Areas and towns/settlements from federal VMAP 1:1,000,000 data.
- Accessed Lands: Any 1 km^2 grid cell that contains known access linear development features (originally resolved at 1 km^2).

METHODOLOGY

DAMS: Overlapping large/small hydro stations were removed by using small generating station data for powerplants less than 20 megawatts, and large generating station data for powerplants greater than 20 megawatts. The small hydro station data contains some residual dams that are no longer associated with active hydroelectric generation. Any overlapping (within 2 km) 1:2 million dams coinciding with the generating station data were visually removed from the database before merging with the generating station data.

MINES: 1997 federal mines dataset. This was the most consistent data for the country, and focused on more industries than many of the provincial datasets that are based on mineral occurrence.

ABANDONED MINES: provincial and some federal (1990) mineral occurrence data that contains data on mine deposits which were "past producers". The resolution and completeness of data is likely to vary across provinces.

SETTLEMENTS: Approximately 1990 1:1,000,000 Digital Chart of the World settlements and built-up areas.

WATERSHEDS: The analysis focuses on "Forested Watersheds", which are watersheds that overlap Rowe forested Forest Regions. This equates to watersheds south of the tree line, and excludes any watersheds that are exclusively grassland. Watersheds are tertiary drainages as defined by the Water Survey of Canada.

Some coastal/boundary point data did not coincide with the 1:1,000,000 watershed boundaries (off the coast or border). This was minimal compared to the entire number of development points.

MAP 3. CONVERTED AND ACCESSED FORESTS

DEFINITIONS

CONVERTED FOREST LAND: Lands that have been converted from a historic forest/forest ecosystem condition to non-forest land. Grassland conversion is not shown or considered. Subclasses include: Urban (Lands classified as "Developed - Urban/Built Up Area" from 1995 Land Cover of Canada (LC95) data) and Rural (Lands classified as "Developed - Cropland", and "Developed - Cropland/Other" from LC95 data)

ACCESSED FOREST LAND: Lands that are within one kilometer of a known access corridor. Access corridors include roads, trails, railways, pipelines (oil/gas), hydroelectric and telephone transmission lines, seismic lines, and known motorized backcountry routes (e.g. ski-doo trails).

UNACCESSED FOREST LANDS: Lands that are further than one kilometer from a known access corridor. Subclasses are defined according to the Canadian Center for Remote Sensing 1995 Land Cover of Canada (LC95) classes as follows:

Forests: All forest classes, plus "Burns" classes. (LC95 classes 1-12)

NONFOREST LAND: All "Open Land" (tree crown density of less than 10%) classes. (LC95 classes 13,14,15), Grassland (LC95 class 16), all "Barren Land" classes (LC95 classes 17-22), Snow/Ice (LC95 class 31).

WATER: LC95 class 30.

MIXED FOREST LAND: Lands defined as "Developed - Cropland/Woodland" and "Developed - Woodland/Cropland" (LC95 23-26) from LC95 data.

METHODOLOGY

Converted land was approximated by overlaying present urban and rural lands as defined by the LC95 data (1 km^2 resolution) with Rowe Forest Regions.

Access data (transportation and utility corridors) was obtained by province/territory from federal, provincial and territorial governments at the highest obtainable quality (resolution, currency and completeness) within the SOF project budget. These factors (content, quality and cost) vary considerably between provincial/federal agencies themselves, such that input datasets differ in their scale, currency, coverage, features and feature attributes. The analysis resolution was set to 1 km^2 to match the CCRS 1995 Land Cover of Canada data. Lambert Conformal Conic, NAD27, was used as the standard SOF projection for analysis and presentation as this was found to be the most common projection for national datasets.

Due to the variability of data features and attributes across Canada, no attempt was made to differentiate between the different types of access corridors. The impact extent for all linear corridors was modeled as 1 km in width. Input vector data was rasterized using a 1 kilometer resolution grid base on the origin of the Land Cover 1995 data. Grid output cells (1 km^2) were classified as accessed if any input corridor feature was present within a cell. Input raster data was first buffered at its original resolution to a width of 1 km before resampling (nearest neighbor) to 1km resolution. This resulted in minimal loss of combined access features (foreground cells) by area during the resample procedure. This method was verified against the vector rasterization process using Yukon vector data (250m resolution). The twomethods differed in total access area by less than 0.1%.

A resulting access corridor grid was produced for each province/territory. Provincial data was trimmed to exclude any data that was not under its respective jurisdiction, and then merged to produce a national access corridor grid. The national access corridor grid was merged with the LC95 land cover data. Any LC95 cell (excluding water) that corresponded to an access grid cell was reclassed as accessed. In addition, all LC95 Developed classes (cropland, cropland/woodland, woodland/cropland, cropland/other, and urban classes—23 to 29) were reclassed as accessed. This assumes that these classes would be accessible by nature of being developed in whole or in part.

The resulting accessed/unaccessed land cover grid was merged with a national 1:1,000,000 populated places dataset to ensure that small fly-in northern communities were considered within the analysis.

Problems with the analysis may include underestimating accessed forest land due to relatively old provincial inventories for transportation themes (estimated average age of data of 10-20 years); missing data (e.g. seismic line data for the Northwest Territories and Saskatchewan); and coarse scale of certain input datasets.

Certain provinces were analyzed using different methods than described above due to the nature of available data. These include British Columbia for which 1989 BC Ministry of Forests Recreation Opportunity Spectrum (ROS) roadless area analysis was used to determine accessed land. Quebec access was obtained by classifying 1994 satellite imagery (125m resolution. The image data (.tiff format) was imported into ERDAS IMAGINE software as a RGB file. 50 initial classes were reduced to water, forest, burns, and nonforest areas using an unsupervised classification. Nonforest areas include urban, rural, recently cut, and regenerating forest land. Nonforest land within tenured forest management areas were reclassified as accessed land. Alberta (AB) and Northeast BC: These areas were analyzed for linear disturbance density using 1:20,000 to 1:50,000 input data. Lands with densities greater than zero were reclassified as accessed. Data was resampled to 1 km2 resolution during the merge operation with the land/forest cover dataset.

The sharp distinction between accessed land in Alberta and certain unaccessed land along the AB/Saskatchewan (SK) and AB/NEBC/North West Territories (NWT) borders reflects the lack of seismic line data for both SK and NWT. Accessed lands within Newfoundland are likely greatly underestimated due to limited input data (1:1,000,000 scale).

MAP 4. ACCESS DENSITIES IN THE WESTERN CANADIAN SEDIMENTARY BASIN

DEFINITIONS

ACCESSED DENSITIES: Amount of roads and other linear features per area (km/km^2).

METHODOLOGY

Western Canadian Sedimentary Basin: Paper map was scanned to 200 dpi and then ArcInfo used to convert image to grid. Features digitized in Arctools.

Access Densities were created from four different sources as noted in data sources. Data from the National Energy Board of Canada was manually digitized. Where no digital data existed for Alberta, or where data was older than 1995, 1:40,000 aerial photographs and/or 1998 Landsat 5 images were manually digitized. Vector coverages were converted to 50 meter resolution grids using ArcInfo command Arcgrid. Grid code 1 represented a cell with a linear feature while grid code 0 represented a cell without a linear feature. Density of linear features was calculated using the ArcInfo grid command focalsum using a circular window with a radius of 18 cells. The raw density values were reclassified using the ArcInfo command reclassify into seven ordinal categories.

MAP 5. LARGE REMAINING UNFRAGMENTED FOREST AREAS

DEFINITIONS

UNACCESSED FOREST: Contiguous tracts of forest that are further than 1 kilometer from an access corridor. Unaccessed forest blocks are subdivided into the following size categories:

- 200 km^2 to 500 km^2
- 500 km^2 to 10,000 km^2
- Over 10,000 km^2

METHODOLOGY

Using the Unaccessed forests theme derived from the converted/unaccessed analysis, unaccessed forest grid cells were grouped into regions of contiguous forest using the Arc/Info regiongroup command. These regions were then classified into size categories as listed above. Final unaccessed forest blocks were combined with the 1995 Land Cover and Rowe Forest Region (1:7.5 million) data to produce summary spreadsheets by Rowe Forest Region. Due to the variability of input data quality, and the coarse (1 km^2) analysis resolution, forest blocks less than 100 km^2 were ignored within summary results.

MAP 6. BRITISH COLUMBIA'S RAINFOREST: FOREST DEVELOPMENT PLANS, 1998-2002

Please contact British Columbia Forest Watch or Sierra Club of British Columbia for more information.

MAP 7. ECOLOGICAL LIMITATIONS TO COMMERCIAL FORESTRY

DEFINITIONS

APPROXIMATE COMMERCIAL FOREST ZONE OF CANADA: Approximate boundary for combined (known) area-based forest tenures and forest administrative areas where volume-based tenures or permits are issued.

LIMIT OF CANADA LAND INVENTORY: Limits of combined Canada Land Inventory (CLI) and Ontario Timber Inventory (identical data classification as CLI)

- No or Slight: Lands Having No or Slight Ecological Limitations to Commercial Forestry. Includes CLI Land Capability for Forestry classes 1 and 2.
- Moderate: Land Having Moderate Ecological Limitations to Commercial Forestry. Includes CLI Land Capability for Forestry classes 3 and 4.
- Land Having Severe Ecological Limitations to Commercial Forestry: Includes CLI Land Capability for Forestry classes 5 and 6.

- Severe: Land Having Severe Ecological Limitations That Preclude Commercial Forestry. Includes CLI Land Capability for Forestry class 7.
- No Data: Areas where no digital CLI data was obtainable, or where land was never classified according to forestry capability.
- Protected Areas: World Wildlife Fund "Designated Areas" where industrial activity is precluded within protected areas.

METHODOLOGY

- CLI polygons are represented using the predominant (by area) CLI land capability for forestry class for each polygon. This is similar to the National Atlas Map MCR 4079 Land Capability for Forestry.
- Except for British Columbia, CLI vector data (by 1:250,000 mapsheet) was rasterized at a 250m resolution using predominant CLI land capability for forestry class for grid values. Individual grids were merged and resampled to 1 km^2 for display. Analysis was conducted using merged 250m resolution grid.
- BC data was digitized from 2 1:1,000,000 maps (north and south) using 5 categories described above, and rasterized to 1 km resolution grid.

MAP 8. OPERATING AREAS OF CANADA'S LARGEST FORESTRY COMPANIES

DEFINITIONS

APPROXIMATE COMMERCIAL FOREST ZONE OF CANADA: Approximate boundary for combined (known) area-based forest tenures and forest administrative areas where volume-based tenures or permits are issued.

AREAS OF OPERATION OF CANADA'S LARGEST FORESTRY COMPANIES: The sum of company-controlled area-based tenures plus part of administrative areas in proportion to the volume of cut within the administrative area allocated to the company.

METHODOLOGY

- Tenure data is based largely on World Wildlife Fund's (WWF) assessment of forest tenures in Canada. Licencee information is included with this dataset where known.
- The WWF tenure data was refined by removing WWF Designated Areas (protected areas that preclude industrial development), and lakes greater than 1,000 km^2, from the spatial polygons. Tenure data was updated to mid-1999 for Quebec and British Columbia tenures.

Company operating areas shown on the map include area-based tenures and administrative areas within which companies have volume-based tenure(s).

Forest companies may actually conduct forest operations on only a portion of the land base identified as their operating areas on the map, due to operability constraints and/or the terms of their forest tenure.

MAP 9. CORPORATE CONCENTRATION WITHIN BRITISH COLUMBIA

DEFINITIONS

PERCENTAGE OF ANNUAL ALLOWABLE CUT HELD BY TEN LARGEST FORESTRY COMPANIES: Proportion of the total allocated annual allowable cut (AAC) held by the ten largest British Columbia forestry companies, by volume, within each Ministry of Forests administration unit (Timber Supply Area or Tree Farm Licence). Each unit was displayed by percentage as follows:

- 0 to 33 percent
- 33 to 66 percent
- 66 to 100 percent

PRIVATE LAND: Large areas of private land outside of the British Columbia Ministry of Forests jurisdiction. However, in many cases (notably Vancouver Island), this land is owned and logged by forestry companies.

PROTECTED AREAS: World Wildlife Fund "Designated Areas" where industrial activity is precluded within protected areas.

METHODOLOGY

- Individual company logging allocations and total annual allowable cut volumes by Ministry of Forests administrative unit (Timber Supply Area, Tree Farm Licence) were obtained from the David Suzuki Foundation, Vancouver, BC. These figures are based on February 1998 Ministry of Forests data.
- The Ministry of Forests Small Business Forest Enterprise Program (SBFEP) was not included as a single corporate entity due to its allocation to numerous smaller companies.
- Corporate ownership was updated for major companies to December, 1999, to reflect recent mergers and acquisitions within the British Columbia forest industry.
- The ten largest British Columbia forestry companies were determined by totaling volume allocations within all Timber Supply Areas and Tree Farm Licences.
- The percentage of annual allowable cut held by ten largest forestry companies was determined by dividing their combined allocated volume by the total annual allowable cut (including SBFEP) per administrative unit.

MAP 10. FIRST NATIONS, MÉTIS, AND INUIT LAND CLAIMS AND TREATIES

DEFINITIONS

UNSETTLED LAND CLAIMS: Major Aboriginal comprehensive land claims based on the assertion of continuing Aboriginal title to lands and resources that have not been addressed by treaty or through other legal means. Claims are based on the Department of Indian and Northern Affairs (INAC) 1999 listing of comprehensive claims only, and is therefore not considered exhaustive. (see *Comprehensive Claims Policy and Status of Claims*, www.inac.gc.ca/subject/claims/comp/briem.html)

MODERN LAND CLAIM AGREEMENT: Settled Aboriginal land claims since the inception of the federal comprehensive claims policy in 1973. Such claims normally include full ownership of certain lands in the area covered by the settlement, plus wildlife harvesting rights, participation in resource management, and some form of self-government within the larger settlement area.

HISTORIC TREATIES: Major historical treaties established between First Nations and the Crown prior to 1973.

FIRST NATIONS, MÉTIS AND INUIT COMMUNITIES: Includes Indian Reserves, Settlements, and "Other Communities" (communities with significant populations of aboriginals) as defined by INAC.

EXISTING FOREST COVER: Forests as defined in Map 3, excluding cropland/woodland.

METHODOLOGY

- Treaty outlines and/or location from National Atlas Map 4162 *Indian Treaties*. Digital data obtained from Geogratis, NRCAN. All boundaries are approximate only.
- Unsettled land claims and land claim settlements compiled from INAC and provincial aboriginal agencies. All boundaries are approximate only. Where no boundaries were obtained, point locations were approximated.
- In several provinces and territories, (British Columbia, Yukon Territory, Northwest Territory, Newfoundland and Labrador, Quebec) individual land claims overlap and cover most of the landbase. These have been shown collectively using jurisdictional boundaries only.

APPENDIX 2. THE REVIEW PROCESS

A key principle of Global Forest Watch is the firm belief that transparency and accountability are essential for the development of better natural resources management. In preparing this report, we faced difficulties both in compiling existing information and ensuring that our datasets were consistent for use in national-level analyses. In the interest of promoting open, public, and transparent information policies, GFW products include detailed notes on data we have assembled (see Appendix 1) and a summary of the major comments experts provided in reviewing early drafts. These comments are listed below, along with details on how comments were addressed. A more comprehensive set of comments are available on our website, including full sets of comments from many of the listed groups and individuals.

THE REVIEW PROCESS

This report and the accompanying maps underwent a detailed review process involving both World Resources and GFW Canada partners and external reviewers.

The draft report was sent to over 60 reviewers at the end of October 1999. We received feedback from over 30 people and organizations. Reviewers included representatives from government, industry, academia, and environmental groups.

Government: Canadian Forest Service (coordinated through Claude Leger), Peter Hall (CFS), B.C. Ministry of Forestry (Chief Forester Larry Pederson, David Morel, and other staff), Frank Ahern (Canadian Centre for Remote Sensing).

Industry: Alberta Pacific Industries (Bill Hunter), Weyerhauser (Jean-Pierre Martel and Linda Coady), Western Forest Products, West Fraser Timber, International Forest Products, Fletcher Challenge Canada, Canadian Forest Products, and TimberWest Forest Limited.

Academics: William Pruitt (University of Manitoba Department of Zoology), Mark Harmon (University of Oregon), Patricia Marchak (University of Victoria), Monique Ross (Canadian Institute of Resource Law, University of Calgary), Peter Duinker (Dalhousie University), Erik Kasishcke (Geography Department, University of Maryland), and Ajit K. Krishnaswamy (International Institute for Sustainable Development).

Other Forestry and Ecology Experts: Trevor Jones, Brad Stelfox, Jim Ball (former CFS), and Peggy Smith (National Aboriginal Forestry Association).

ENGOs: Arlin Hackman and Tony Iacobelli (World Wildlife Fund Canada), Martin Von Mirbach (Centre for Forest and Environmental Studies), Matt Price (Natural Resource Defence Council)

Several WRI staff also provided input: Nels Johnson, Nancy Kete, Lars Laestadius, Peter Leimgruber, Marta Miranda, Cathy Plume, Mark Rowheder, Nigel Sizer, Tony Janetos, and Andrew Malk.

In addition, Map 2 was circulated to the following individuals: Jim Ball, Peter Duinker, Tony Iacobelli, and William Pruitt.

MAJOR REVIEW COMMENTS, AND HOW THEY WERE ADDRESSED

Most of the comments received during the review process concerned the structure and presentation of materials within this report as well as our use and interpretation of certain datasets. In reviewing initial drafts, some individuals questioned whether materials lived up to GFW's mandate of presenting balanced, objective information. They felt the overall report was implicitly critical of forest management practices in Canada. Other reviewers, however, felt that we presented too rosy a picture of the status of Canada's forests.

- Structure of the report not clear and messages difficult to find

Several individuals noted that no framework was provided with the first draft and that it was unclear what the report was trying to accomplish. We have addressed this by adding an introduction that provides a road map for the report. We have also restructured and simplified initial drafts of this report, so that results could be more clearly communicated to our audiences. Our report follows the outline of other GFW reports, which focus on four themes: forest trends (extent, condition), development trends, key actors, and performance (legislation and compliance). Thus, our report provides an overview of development trends in the forests, although we provide a more in-depth look at the forest industry.

- Selective use of data and negative interpretation

The review process was very helpful in identifying areas where we needed to provide a more balanced presentation. We cut much of the text and let the data speak for themselves; we revised what was left in an effort to achieve a balanced tone. We revisited data sources to ensure that we addressed concerns about selective reporting and interpretation of data. We have also added sub-sections on various items to ensure a more well-rounded picture of the forest industry and the legislative and policy context. For example, we added materials on natural disturbance regimes to provide a more complete picture of forest management issues. We have also provided more examples of positive steps in forest management and specific references to processes and initiatives recommended by reviewers, such as certification processes. We have also revised our presentation of materials on clearcutting and added a section on Allowable Annual Cuts.

Some reviewers commented that our initial "Are Canada's Forests at Risk?" section implied that development was inherently bad. We reworded this section to present development trends and then analyzed them against a series of values people derive from forests (at risk depends on what you care about). We have also revised the jobs and economic benefits section to improve its accuracy and include the most recent government data.

- Indicators failed to incorporate information on the relative impacts of activities

Several reviewers noted that our indicators failed to incorporate information on the relative impacts of different development activities. For example, the environmental impacts of oil sands development versus roads, or the varying levels of traffic and clearing associated with different types of access routes. As noted in the text, we were unable to factor this in to a national-level analysis given existing datasets.

- Report is largely a rehash of existing data

Some reviewers felt that we were simply providing a compendium of existing data rather than a systematic assessment of forestry issues. Others felt we were undertaking original analysis. We have drawn much more heavily on our maps and analysis of digital data in this version. Our initial work involved collecting and synthesizing data and presenting these (where possible) in map format. We also present different analysis of some core government data on the forest sector, including forest fires and sustainability of harvest rates.

- The report contains judgmental, advocacy language and does not reflect the nonadvocacy nature of GFW

Several reviewers flagged draft materials they deemed judgmental, or otherwise questioned GFW's commitment to presenting objective, balanced information. We edited or deleted those sections of the text accordingly, and provided additional references to support poorly substantiated materials. Some reviewers felt quite strongly that our use of some terminology such as development and logging was not strong enough, while others felt we used these terms in a negative sense.

- Request for more in-depth treatment of certain development-related topics

Many reviewers recommended we deal with specific development-related topics in greater depth. These included, for example, a more thorough treatment of the role of aboriginals in the forest sector and comparative analysis of the impacts of different forest management regimes. Due to data and time limitations, we were unable to provide analysis at the level of detail we would have liked. Global Forest Watch Canada will be addressing these and other issues through subsequent monitoring and reporting activities.

- Maps do not always accurately present data:

Some reviewers had useful comments on the quality and nature of the maps.

Converted and Accessed Forest Map: The main concerns with the original access map related to the inclusion of oil and gas seismic exploration lines within the category of linear features used to determine accessed forest. We have attempted to clean up these data layers. Due to the nature of the datasets, as discussed in Appendix 1, there is great variability in the inclusion of linear features in the various data sets. We have made minor changes to the underlying data layers and have improved presentation to help improve the new combined access/conversion map. We will provide a new version of this map by March as part of our next steps to provide accurate, up-to-date information.

Ecological Limitations to Commercial Forestry: The main concern with this map was that the underlying Canada Land Inventory data is 30 years old and that our commercial forest zone, which is a combination of all types of existing forest management areas, overestimated the area that will be logged. We have addressed these concerns by comparing the CLI to current British Columbia productivity data to verify how accurate the CLI data is. The major differences were in categories of what is deemed productive. We have revised our analysis of the meaning of our map to reflect these findings. As well, we have clarified that our figures do not imply that all areas within the commercial forest zone will be harvested.

Operating Areas of Top Companies in Canada: Several reviewers complained that the map was an inaccurate representation of the control of forest harvesting and of companies working in volume-based harvesting areas. We renamed the map from "Who Controls Canada's Forests" to "Areas of Operation of Top Companies in Canada." We also redid our analysis so that we factored in all available forest harvesting agreements, including licences in British Columbia. We dealt with the issue of accurate portrayal of volume-based harvesting rights (which are generally held by more than one company) by providing a legend and set of notes for which companies appear on the national map. We have also added in a new map showing tenures for British Columbia and the percentage of tenure areas that are held by the largest 10 companies in British Columbia. We will conduct similar analysis for Quebec, the other major volume-based province.

We will continue to improve on our datasets and analysis. Updates will appear on our web-site and comments and suggestions can be provided to us on the site.

APPENDIX 3. POSITIVE AND NEGATIVE ASPECTS OF DEVELOPMENT ACTIVITIES

Hydroelectric Power

EXTENT

- Canada was the number one hydroelectric power generator in the world in 1995.[1]
- Hydropower generated 331,619 gigawatt hours of electricity in 1995.[2]
- There are 618 large dams over 10 meters tall.[3]
- Five large dams generate more than 1,000 megawatts each.[4]
- The headpond of a single large dam in northern Quebec (Le Grande, Phase 1) floods an area of more than 13,500 km^2, almost half of which was originally forested.[5]

SOCIOECONOMIC BENEFITS

- Hydroelectric power is the largest domestic source of electric energy, representing 61 percent of the nation's supply in 1997.[6]
- Compared to coal, oil, or natural gas, hydroelectric power releases relatively low levels of air pollutants, carbon dioxide, and other greenhouse gases.

ECOLOGICAL IMPACTS[7]

- Dams and reservoirs can harm downstream aquatic communities through loss of nutrients, primary productivity, and reduced streamflow. For example, fish may lose their breeding grounds and feeding areas; as a result, they may not be able to migrate, grow, or spawn.
- Direct flooding from dams contributes to loss of habitat.
- When dams and reservoirs are constructed, they may flood forested peatlands. These floods cause a release of greenhouse gases (CH_4 and CO_2) to the atmosphere. The intensity and duration of the emissions depends on the amount of flooding, the age and location of the reservoir, and the amount of plant biomass.
- By altering habitat, large-scale hydropower development may limit biodiversity. Native species may be unable to quickly adapt to drastic ecosystem alterations and exotic species may take their place.
- Bioaccumulation of methylmercury—a neurotoxic, organic molecule produced by bacteria—can be common in fish. Contamination occurs during reservoir construction and is typically confined to fish in the reservoir and immediately downstream. Humans can also bioaccumulate methylmercury by eating affected fish. Aboriginal communities and areas with subsistence economies are at particular risk. Methylmercury contamination typically lasts at least 20-30 years.

Oil and Gas and Oil Sands Development

EXTENT

- Canada is the third largest producer of natural gas in the world.[8]
- Estimated recoverable gas reserves exceed 150 trillion cubic feet.[9]
- In 1997, gas output was 5.5 trillion cubic feet.[10]
- Gas reserves are largely confined to the Western Canadian Sedimentary Basin in Alberta, Saskatchewan, and northeastern British Columbia; 82 percent of supply is from Alberta.[11]
- Four major oil sands deposits (Athabaska, Peace River, Wabasca, and Cold Lake) contain an estimated 600 million cubic meters of bitumen, the world's second largest known source of oil. These deposits are equivalent to 300 billion barrels of crude oil.[12]
- Two large oil sands companies operate in the boreal region—Suncor and Syncrude. An additional five oil companies plan to spend $25 billion (US $17 billion) over the next 25 years to develop additional oil sands projects.[13]
- Approximately 89,000 oil and gas well sites were located in Alberta's boreal forest by 1997.[14]

SOCIOECONOMIC BENEFITS

- Oil and gas contributed $18.6 billion (US $13 billion) or 2.4 percent to Canada's GDP in 1998.[15]
- Oil and gas industry payments to governments totaled $8 billion (US $5.5 billion) in 1997.[16]
- The crude oil and gas production industry generated 83,000 direct jobs in 1997.[17]

ECOLOGICAL IMPACTS

- Main impacts of seismic lines, trails, and pipelines on wildlife are individual species disruption, habitat disruption, direct mortality (hunting), and indirect mortality (from increased predator/prey contact).[18]
- Old seismic lines may be used as travel routes, which benefit some wildlife. Recolonizing vegetation may provide additional food sources. Habitat advantages for prey species may be offset by increased predation.[19]
- Treatment of bitumen from oil sands can result in contamination of surface waters, aquifers, soils, and the atmosphere due to high sulphur, nitrogen, and trace minerals content.[20]

Mineral Exploration and Mine Development

EXTENT

- 44 million hectares of mineral claims were registered in 1997. Claims increased by 232 percent over 1996, largely due to increased interest in potential diamond properties.[21]
- 675 nonpetroleum mineral exploration companies operated in 1997.[22]
- In 1998, 293 metal, nonmetal and coal mines, as well as 3,000 stone quarries and sand and gravel pits, were in operation.[23]
- Over 20 types of metals (e.g. copper, nickel), 22 nonmetals (e.g. salt, sulphur), 5 structural materials (e.g. lime, clay), and 4 mineral fuels (e.g. coal, crude oil) are produced in Canada.[24]

SOCIOECONOMIC BENEFITS

- Mining contributed $26.4 billion (US $18 billion) or 3.7 percent to GDP in 1998.[25]
- Mining directly employed 367,000 people in 1998.[26]

ECOLOGICAL IMPACTS[27]

Exploration

- Activities: airborne and ground-based geochemical and geophysical surveys; prospecting; claim staking; line cutting; stripping; drilling and trenching; road/trail building; and bulk sampling.
- Environmental issues (subject to mitigation/prevention measures): land removed from protection options; trail/road trenching and erosion; and habitat disruption.

Mining and Drilling

- Activities: stripping/storing of soil and vegetation overburden; ore extraction; crushing/grinding, flotation or chemical concentration of ore; mine and surface water treatment, storage of waste rock and tailings; and processing of mineral concentrate.

Environmental issues: wildlife and fisheries habitat loss; sedimentation in surface waters; containment of toxins in tailings ponds; acid generation from waste rock and pit walls; heavy metal leaching from acid mine drainage; and wind-borne dust.

Mine Closure

- Activities: recontouring of pit walls and waste dumps; covering reactive tailings dumps; decommissioning of roads; re-seeding/planting of disturbed areas; and ongoing monitoring and possible water quality treatment
- Environmental issues: seepage of toxic solutions into groundwater and surface water; contamination from acid mine drainage; revegetation failure; wind-borne dust; and slope and tailing impoundment failure.

Roads

EXTENT

- Total estimated length of roads in 1996: 912,200 kilometers, including 246,400 kilometers of paved roads (16,600 kilometers of expressways) and 665,800 kilometers of unpaved roads.[28]
- Total spending on equipment (e.g. cars) and maintenance of roads was $18.7 million (US $13 million) in 1996.[29]

SOCIOECONOMIC BENEFITS

- In 1997, approximately 300,000 people were employed in the truck, bus, and urban transit sectors. In addition, just under 70,000 people were employed in jobs related to highway infrastructure.[30]
- Primary roads provide rural communities with better access to employment, shopping, education, and medical care.

- Secondary roads enable economic access to timber, minerals, and other natural resources.
- Roads facilitate tourism and recreational use of forest areas.

ECOLOGICAL IMPACTS

- Impact on wildlife includes individual species disruption, habitat loss and fragmentation, direct mortality (hunting and road kill), and indirect mortality (from increased predator/prey contact).[31]
- Many wildlife populations decline as road densities increase.[32] Large mammals, such as wolf, grizzly bears, elk, and caribou are the most adversely affected.[33]
- Physical impacts of roads include accelerating erosion from road surfaces; alteration of surface water flows and the timing of peakflows; erosion during flood events; increased landslides; and loss of soil productivity.[34]
- For aquatic habitat and fish populations, roads may introduce barriers to migration, cause changes in water temperature, and alter streamflow regimes.[35]
- Roads, especially closed roads or roads with little traffic, may be used as travel routes benefiting some wildlife. Recolonizing vegetation may provide additional food sources. Habitat advantages for prey species may be offset by increased predation.[36]

1 Natural Resources Canada. "Natural Resources Canada Energy Factsheets: Electricity." Online at: http://nrcan.gc.ca:80/apec/nr-elec.htm. (January 17, 2000).

2 Natural Resources Canada. "Natural Resources Canada Energy Factsheets: Electricity." Online at: http://nrcan.gc.ca:80/apec/nr-elec.htm. (January 17, 2000).

3 Environment Canada, *A Primer on Fresh Water* (Ottawa: Minister of Supply and Services, 1994) Online at: http://www.ec.gc.ca/water/en/info/pubs/primer/e_prim03.htm#a4 (January 18, 2000).

4 D.M Rosenberg et al. 1987. "The environmental assessment of hydroelectric impoundments and diversions in Canada," in "Canadian Aquatic Resources," M.C. Healey and R.R. Wallace, eds. *Can. Bull. Fish. Aquat. Sci.* 215: 71-104. (Cited in D.M. Rosenberg et al. 1997. "Large-scale impacts of hydroelectric development," *Environmental Review* 5: 27-54).

5 Rougerie, J-F. "James Bay development project: Hydroelectric development in northwestern Quebec." *Can. Water Watch* 3: 56-58. (Cited in D.M. Rosenberg et al. 1997. "Large-scale impacts of hydroelectric development." *Environmental Review* 5: 27-54).

6 Canada's Electricity Sector. Online at: http://www.energy.ca/ELECTRIC.html (January 17, 2000).

7 D.M. Rosenberg et al. 1997. "Large-scale impacts of hydroelectric development." *Environmental Review* 5: 27-54.

8 Canadian Association of Petroleum Producers. Online at: http://www.capp.ca/02cxb.html (January 14, 2000).

9 Canada's Natural Gas Sector. Online at: http://www.energy.ca/GAS.html (January 14, 2000).

10 Canada's Natural Gas Sector. Online at: http://www.energy.ca/GAS.html (January 14, 2000).

11 Canada's Natural Gas Sector. Online at: http://www.energy.ca/GAS.html (January 14, 2000).

12 Alberta Environmental Protection. *The Final Frontier: Protecting Landscape and Biological Diversity within Alberta's Boreal Forest Natural Region, Protected Areas Report #13*, (Edmonton: Alberta Environmental Protection, 1998), p. 86.

13 Alberta Environmental Protection. *The Final Frontier: Protecting Landscape and Biological Diversity within Alberta's Boreal Forest Natural Region, Protected Areas Report #13*, (Edmonton: Alberta Environmental Protection, 1998), p. 86.

14 Alberta Environmental Protection. *The Final Frontier: Protecting Landscape and Biological Diversity within Alberta's Boreal Forest Natural Region, Protected Areas Report #13*, (Edmonton: Alberta Environmental Protection, 1998), pp. 78, 81.

15 Natural Resources Canada. Online at: http://www.nrcan.gc.ca/statistics/factsheet.htm (January 14, 2000).

16 Canadian Association of Petroleum Producers. Online at: http://www.capp.ca/02cxb.html (January 14, 2000).

17 Canadian Association of Petroleum Producers. Online at: http://www.capp.ca/02cxb.html (January 14, 2000).

18 Arc Wildlife Services, *The Effects of Linear Development on Wildlife: A Review of Selected Scientific Literature, Pub #1998-0002* (Calgary: Environmental Research Advisory Council, 1998), pp. 20-22.

19 Arc Wildlife Services, *The Effects of Linear Development on Wildlife: A Review of Selected Scientific Literature, Pub #1998-0002* (Calgary: Environmental Research Advisory Council, 1998), p. 22.

20 Alberta Environmental Protection, *The Final Frontier: Protecting Landscape and Biological Diversity within Alberta's Boreal Forest Natural Region, Protected Areas Report #13*, (Edmonton: Alberta Environmental Protection, 1998), pp. 87-90.

21 Ginette Bouchard, "Mineral Exploration Activity in Canada," in *Canadian Minerals Yearbook 1998*, (Ottawa: Natural Resources Canada-Minerals and Metals, 1999), p 3.5. Online at: http://www.nrcan.gc.ca/mms/cmy/CMY_E2.html (January 17, 2000).

22 Natural Resources Canada, Minerals and Metals Sector. Online at: http://www.nrcan.gc.ca/mms/efab/mmsd/facts/canada.htm (January 17, 2000).

23 Natural Resources Canada, Minerals and Metals Sector. Online at: http://www.nrcan.gc.ca/mms/efab/mmsd/facts/canada.htm (January 17, 2000).

[24] Minerals and Mining Statistical Division. "Statistical Report," in *Canadian Minerals Yearbook 1998*, (Ottawa: Natural Resources Canada, Minerals and Metals, 1999), p. 64.4. Online at: http://www.nrcan.gc.ca/mms/cmy/CMY_E2.html (January 17, 2000).

[25] Natural Resources Canada, Minerals and Metals Sector. Online at: http://www.nrcan.gc.ca/mms/efab/mmsd/facts/canada.htm (January 17, 2000).

[26] Natural Resources Canada,Minerals and Metals Sector. Online at: http://www.nrcan.gc.ca/mms/efab/mmsd/facts/canada.htm (January 17, 2000).

[27] Environmental Mining Council of British Columbia, *More Precious than Gold . . . Mineral Development and the Protection of Biological Diversity in Canada* (Victoria: Environmental Mining Council of BC, 1998). Online at: http://www.miningwatch.org/emcbc (January 17, 2000).

[28] Central Intelligence Agency, *The World Factbook 1999* (Washington, DC: CIA, 1999). Online at: http://www.odci.gov/cia/publications/factbook/index.html (January 17, 2000).

[29] Transport Canada, *Transportation in Canada 1998 Annual Report* (Ottawa: Minister of Public Works and Government Services, 1999), p. 7. Online at: http://www.tc.gc.ca/pol/en/ANRE1998/options.htm (January 18, 2000).

[30] Transport Canada, *Transportation in Canada 1998 Annual Report* (Ottawa: Minister of Public Works and Government Services, 1999) p. 88. Online at: http://www.tc.gc.ca/pol/en/ANRE1998/options.htm (January 18, 2000).

[31] Arc Wildlife Services, *The Effects of Linear Development on Wildlife: A Review of Selected Scientific Literature, Pub #1998-0002* (Calgary: Environmental Research Advisory Council, 1998), pp. 16-20.

[32] United States Department of Agriculture-Forest Service, *Forest Service Roads: A Synthesis of Scientific Information* (Washington, DC: USDA-Forest Service, 1999), p. 49. Online at: http://www.fs.fed.us/news/roads (January 17, 2000).

[33] Silva Forest Foundation, *Assessing the Ecological Impacts of Timber Harvesting: Apparent Impacts, Actual Impacts, and Precautionary Forest Development Planning* (Winlaw, BC: Silva Forest Foundation, 1999), p. 8. Online at http:/www.silvafor.org/docs/docs.htm#buffer (January 17, 2000).

[34] United States Department of Agriculture-Forest Service, *Forest Service Roads: A Synthesis of Scientific Information* (Washington, DC: USDA-Forest Service, 1999), pp. 21-24. Online at: http://www.fs.fed.us/news/roads (January 17, 2000).

[35] United States Department of Agriculture-Forest Service, *Forest Service Roads: A Synthesis of Scientific Information* (Washington, DC: USDA-Forest Service, 1999), pp. 35-37. Online at: http://www.fs.fed.us/news/roads (January 17, 2000).

[36] Arc Wildlife Services, *The Effects of Linear Development on Wildlife: A Review of Selected Scientific Literature, Pub #1998-0002* (Calgary: Environmental Research Advisory Council, 1998), p. 19.